ARDUINO™ ROBOT BONANZA

ABOUT THE AUTHOR

Gordon McComb has written 65 books and thousands of magazine articles—over a million copies of his books are in print, in more than a dozen languages. For 13 years, Gordon wrote a weekly syndicated newspaper column on computers and high technology, which reached several million readers worldwide. He's a regular contributor to *SERVO Magazine* and other publications, and maintains an active Web site dedicated to teaching the art of science of robot building. He is the author of the best-selling *Robot Builder's Bonanza*, now in its fourth edition.

ARDUINO™ ROBOT BONANZA

GORDON McCOMB

New York Chicago San Francisco Lisbon London Madrid
Mexico City Milan New Delhi San Juan Seoul
Singapore Sydney Toronto

Cataloging-in-Publication Data is on file with the Library of Congress

Arduino™ Robot Bonanza

1 2 3 4 5 6 7 8 9 0   DOC DOC   1 0 9 8 7 6 5 4 3

ISBN   978-0-07-178277-7
MHID   0-07-178277-X

Sponsoring Editor Roger Stewart	**Copy Editor** Lisa McCoy	**Composition** Cenveo Publisher Services
Editorial Supervisor Patty Mon	**Proofreader** Paul Tyler	**Illustration** Cenveo Publisher Services
Project Manager Tania Andrabi, Cenveo® Publisher Services	**Indexer** Ted Laux	**Art Director, Cover** Jeff Weeks
Acquisitions Coordinator Molly Wyand	**Production Supervisor** Jean Bodeaux	

For my mother,
Edna Kathryn Lane Forbes
1920–2012

CONTENTS

Part 3—Hands-on Arduino Robot Projects

ACKNOWLEDGMENTS

Special thanks go to Ken Gracey, Matt Gilliland, Jim Carey, and my many friends at Parallax; to Jim Frye of Lynxmotion; to Nathan Seidle, Pete Dokter, and Robert Cowan of SparkFun; to Gerry Coe of Devantech; to Russell Cameron, Claudia, and the crew at DAGU; to Roger Stewart and the editors at McGraw-Hill Education; and to my agent Matt Wagner.

And as always, to my wife Jennifer.

INTRODUCTION

THIS IS YOUR ROBOT ... WITH AN ARDUINO BRAIN

When I first started building robots, the microprocessor had not yet been invented. Robot brains were limited to hand-wired boards filled with gompy resistors, super-expensive transistors, and maybe even a relay or two.

Today we have *microcontrollers*, wonderful micro-miniaturized wonders that combine a computer with the ability to directly connect to lights, alarms, motors, solenoids, sensors, and other physical things. In fact, this whole concept now goes by the catchphrase *physical computing*. I mention this because physical computing plays directly into building robots. That's why microcontrollers like the Arduino are so important in robotics.

The Arduino, first designed to help student designers integrate electronics and mechanics into their work, is one of the most popular microcontrollers now available. It's meant as a tool for building projects that sense and control their nearby world.

The Arduino is like a brain in a jar ... by itself it lacks the ability to detect its environment or manipulate anything. It's up to you to hook up your choice of sensors, motors, and other control electronics, then program the Arduino to do your bidding. And here's where this book comes in: *Arduino Robot Bonanza* helps you build a half-dozen robots that use the Arduino as a central brain. It shows you how to wire things up, then write programs so your 'bot follows your commands.

The projects in *Arduino Robot Bonanza* are designed to be reproducible using ordinary shop tools and average construction skills. There's no cutting heavy metal or welding, and each of the six robots is constructed using parts that are commonly available. None of the projects rely on hard-to-find or surplus components. While robot building can be an expensive endeavor, all plans in this book favor lower-cost parts, and encourage reuse.

FREE ONLINE CONTENT, PARTS FINDER, AND BONUS GOODIES

This book comes with free online content: the *ARB Online Support* site. See Appendix A for the lowdown:

- Visit the *Project Parts Finder* for all the parts for the projects in this book
- Downloadable source code for all projects
- New and updated links to Web sites and manufacturers
- Enhanced and updated robot e-plans
- Bonus articles, tutorials on robot construction, and more

START HERE, BUT DON'T STOP HERE

Think of *Arduino Robot Bonanza* as a way to get somewhere, and not a final destination. The six projects in this book are meant to be springboards—ideas to get you started. The programming code—*sketch* in Arduino parlance—is kept simple so you can more easily review how it works, tear things apart, and add your own special flair.

Many of the projects are provided with at least programming code, printed directly in the book if it's short enough, but always available as a quick and easy download on the ARB Online Support site.

I consider all my robot projects works-in-progress, and I'm always tweaking and adjusting. The ARB Online Support site also provides alternative versions of code, and I welcome your submissions of enhanced and improved sketches!

Now it's time to get in the car, rev up the engine, and start your journey!

Part 1

Arduino Robot Basics

Chapter 1

Introducing the Arduino Development Platform

I built my first robot when I was five years old.

It didn't do much. Truth is, it didn't do anything. It was made from a discarded coffee can, some wires, and a few radio tubes I found in my father's workbench. I sat on the can to flatten it out—that's what robots look like, after all—and stuffed the wires and tubes inside.

Though my first *Metal Man* was just a pile of dusty old junk, it started me on a lifelong path to building bots of all kinds, shapes, and sizes. Until recently, building real robots took a lot of time, money, and technical skill. Even if you could pull it off, the results weren't always impressive. There were hard limits to technology and budget.

Times have changed.

Today, with just a small collection of inexpensive parts, you can construct a fully functional and thinking robot. Wheels, tank treads, legs… your choice. Little or no soldering is required. To make changes you just write a new program. Parts are interchangeable, so you can reuse components from an older robot on a newer one. That saves money.

You might call all this a revolution in amateur robotics. And if it is a revolution, what's behind it? Many things, but here are three important reasons:

- *Inexpensive and widely available microcontrollers,* essentially self-contained computers the size of a postage stamp. See Figure 1-1 for an example—it doesn't look like much, but there's a lot of power under the hood. Microcontrollers are special because they're made to connect with things like blinking lights, speakers, microphones, sensors, motors, and solenoids—you know, the stuff of robots.
- *Easy-to-use Arduino,* a leader among a new and growing cadre of low-cost and easy-to-use microcontroller development platforms. The Arduino was originally designed for use in art and design projects, and was intended to simplify microcontroller programming. Gear heads

Figure 1-1 A microcontroller on a single integrated circuit (IC). Microcontrollers are intended for *physical computing,* for directly connecting to the physical world around them.

are still allowed, of course, because the Arduino does all of this without taking away any power or flexibility.

- *Affordable, reusable building blocks* for attaching components to a microcontroller. Want a robot that seeks out the heat of your pet hamster? There's no need to construct a complex circuit which might not work when you're done. There's a ready-made component for this very thing. Or how about a robot that's commanded by voice control—just say "run" and "stop" and the robot obeys. There's a module for that, too. Or how about a robot that sings its own tunes, a mini orchestra on wheels. Not only is there a module for that, there are several to choose from.

This book takes these three revolutionary ideas and presents not one, but *six* complete robot projects that you can build and program yourself. All the robots are autonomous, meaning they think for themselves, reacting to you and their environment.

Each project starts with construction plans—none of the projects require specialized tools or advanced shop skills. These are bots you can make in your garage with standard materials. You'll see how to wire everything up and demonstrate that everything is working. From there you're free to customize your creations the way you want. There's no such thing as a one-size-fits-all robot!

Why Robotics

Okay, so you've always dreamed of building a robot army to help you take over the world. Noble thinking, but set those plans aside for a moment, and contemplate the other reasons to build robots.

Who hasn't marveled at the bomb-sniffing robots that risk their aluminum guts to save the lives of us people. Or how about those amazing rocket-hopping bots that tour alien landscapes, like the Mars *Sojourner* in Figure 1-2, beaming back pictures of sunsets on other planets.

And if this weren't enough, the art and science of robots goes *far* beyond the robots themselves. These plucky machines are chock full of the latest and greatest technology, stuff that's used in a wide variety of consumer, industrial, and military products. Experimenting with these technologies gives you a front row, face-against-the-screen, close-up look.

Years before most people ever heard of the term *smartphone,* we robo-builders were playing with (and learning about!) all the fun junk that goes in them: electronic compasses, accelerometers, gyroscopes, global positioning satellites, microcontrollers, touch screens, artificial intelligence, voice control, speech synthesis, and more. When building a robot, you get to play with the technologies of today and tomorrow.

Figure 1-2 NASA's Mars *Sojourner* robotic rover, being inspected before being launched by rocket to another world. *Photo courtesy NASA.*

Why the Arduino for Robotics

Robots are more fun when they think for themselves. From the early days robots came to life, thanks to an old clanky relay and some tubes, or a couple of transistors soldered to a hand-constructed circuit board. Other robotic processing units have used bulky computers in one form or another, even the innards of programmable calculators.

These techniques were fine in their day, but modern electronics offers so much more, and for much less money. That's where the microcontroller comes in, a single-chip computer that's made to interface—connect with—devices in the real world. Microcontrollers aren't new; they've been around for several decades. But until fairly recently, they've carried a heavy cost, either in the parts themselves or the requirements needed to program them.

There have been several attempts to simplify microcontrollers for the masses. Example: The Parallax BASIC Stamp, first released in 1992, provided a functional microcontroller on a small circuit board. You used your own computer, connected to the BASIC Stamp by cable, to write programs. Similarly, the LEGO Mindstorms has at its core a microcontroller that you program to read sensors and operate motors.

In 2005, some teachers and students at a design school in Italy devised the *Arduino* as a low-cost alternative to more expensive microcontrollers commonly used for creating

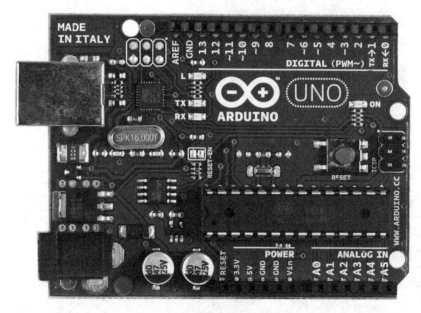

Figure 1-3 The Arduino Uno development board, ready for use in a robotics project. Just connect it to your PC via a USB connection, and upload a program into its memory.

interactive projects. Instead of building everything from scratch, the Arduino's creators leveraged several *open-source* projects, thereby saving them hundreds, if not thousands, of hours of time. It also greatly expanded the resources available to keep development going.

The Arduino, shown in Figure 1-3, is a *mashup* of already-established projects, but its parts are orchestrated to remove many of the complexities common to microcontroller programming. And because the Arduino is an open-source project, it can be freely replicated by others without paying royalties.

You can buy an "official" Arduino board, or you can also build your own from readily available plans. Plus, dozens of independent companies and developers sell Arduino-compatible microcontrollers, giving you plenty of options. This helps to keep prices low.

Robot Stuff You Can Do with the Arduino

So what are some of the *robot-y* things you can do with an Arduino? Here's just a short list, all of which just happen to be actual projects presented in this book:

- *Make your robot detect and immediately react to touch and light.* Your bot can seek out contact and bright light sources, or shy away from them. Simple changes in program code dictate how your robot behaves in response to various kinds of physical stimulation.
- *Operate different kinds of motors, and control their speed.* Under your program, the motors of your robot start and stop in specific patterns, allowing the machine to avoid or steer around obstacles.
- *Have your robot trace a track of black or colored tape.* Race with other robot builders to see who's fastest or stays on course the best.
- *Have your robot play its own music while it's exploring or interacting with you.* You can use pre-recorded music or sounds, or play back MIDI tunes of your own creation.

- *Explore different locomotion techniques,* including running on wheels, tracks, and even slithering over the ground like a snake.
- *Store pre-defined movements for a robotic arm, and play them back.* This technique is used in industrial robots to replicate a complex series of steps, like assemble and weld a door onto a car. (The arm in this book is suitable for moving much lighter things, but the idea is the same.)
- *Combine multiple sensors to detect objects,* even the feather touch of a robotic kiss. (Hint: try it with your cat!)
- *Sense your body motions and transmit those to a robot on the ground.* Turn and the robot turns. Walk and the robot advances forward.
- *Construct a talking animatronic figure* that responds to the people around it. Have it play games, tell jokes, sing songs.
- *Endow your robot with environmental sensors*—temperature, humidity, atmospheric pressure—and have it wander around your house, recording its findings and even beaming them back to you via television link.

Arduino: Under the Hood

The Arduino is more than just a microcontroller chip. It's really a *development platform* that encompasses both hardware and software (see Figure 1-4).

The Arduino hardware is a complete circuit board, consisting of a microcontroller integrated circuit (*IC*), sockets for plugging in wires to connect with the rest of your circuit, some voltage regulators to provide the right kind of power to all the components, and a USB interface to connect the Arduino to your computer. There are several models of Arduino development boards, in a variety of shapes and sizes. More about this in a bit.

The Arduino software provides the means to program the Arduino from your PC. The software combines a text editor where you can write and refine your programs, plus an automated uploader that transfers your finished programs to the Arduino.

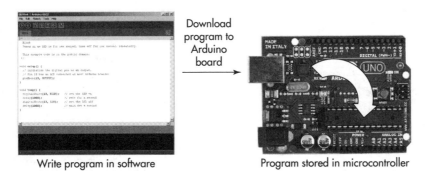

Write program in software   Program stored in microcontroller

Figure 1-4   The Arduino is a crafty combination of both hardware and software, each designed to complement one another while making it easier to program microcontrollers.

A CLOSER LOOK AT ARDUINO HARDWARE

Since its introduction the Arduino has gone through numerous iterations, revisions, and improvements. The parade of different models of Arduino hardware can be a bit confusing at first, but is summarized in the following table. Each type of Arduino has its own best use; you want to match the Arduino hardware with the needs of your project.

	Hardware Class	Representative Model(s)	Features
	Basic	Duemilanove Uno Leonardo	Most common hardware design, provides for up to 19 external connections. This is the most popular Arduino board design. Ideal for: Basic experimenting; main brain on average-size robots.
	Mega	Mega2560 Mega ADK	Expanded connections and more robust microcontroller chip, designed for more demanding tasks. Ideal for: Larger and more sophisticated robots; bot that's connected to a lot of different external devices.
	Mini	Mini Pro Nano	Smaller "stamp-size" footprint, for when you need an Arduino in a tight space. These models tend to lack a USB port, which is used to program the board. You need a USB adapter module to complete the programming connection. Ideal for: Projects where space is at a premium; smallest of robots; robots you wear or carry.
	Breadboard	Boarduino	Designed specifically for use on solderless breadboards. Some cross-over with the *mini* class, but often larger and easier to build if you're constructing one from a kit of parts. Ideal for: Designs where you expect a lot of ongoing changes.

Hardware Class	Representative Model(s)	Features
	Specialty	Lilypad Fio BT Ethernet	Variety of specialty Arduinos designed for specific uses. Examples: The Lilypad is flower shaped with flexible connections, made to be used in clothing and textiles. Ideal for: Various—depends on the board. Lilypad can be sewn into garments (but be careful about tossing it into the washing machine!). Fio and BT are made for wireless applications. Ethernet Arduino is made to connect to Ethernet networks.

FYI The projects in this book demonstrate the use of several of these Arduino hardware classes, but concentrate on the Arduino Uno. In most cases, the designs in this book do not *require* specific types and styles of Arduino hardware, so unless otherwise noted feel free to substitute as needs arise.

Another word for the hardware class is *form factor*, which not only describes the overall size and shape of the Arduino board, but its general physical features, such as the number of external connection points.

The "official" Arduino boards—the ones pictured on the main *www.arduino.cc* Web site— are often referred to as *reference designs*. That simply means they are intended to be used as a reference or starting point for others wishing to develop their own Arduino-compatible hardware. There are numerous reference designs, and each one has spawned numerous derivative Arduino-based boards from other manufacturers and individuals. All this means is that there isn't a single "Arduino board" to choose from.

MAJOR ARDUINO POINTS OF INTEREST

Recall the Arduino hardware contains numerous electronic components, providing a convenient platform for experimenting with different kinds of computer-controlled circuits. See Figure 1-5 as we go down the list of main points of interest on the Arduino tour.

Main Microcontroller

At the heart of the Arduino is the microcontroller itself, a fully self-contained computer specially designed for communicating with external components like lights and motors. The exact size, layout, and model of the microcontroller vary depending on the specific Arduino board.

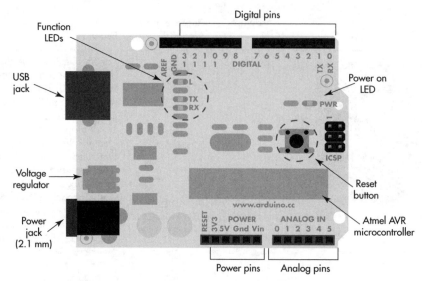

Figure 1-5 The main points of interest on an Arduino Uno board

One commonality is that nearly all are 8-bit controllers manufactured by Atmel, a major provider of microcontrollers for industrial and commercial applications.

● The latest basic class boards, including the Arduino Uno, use the *Atmel AVR ATmega328P* microcontroller. This chip is available in two basic forms: *dual inline package* (DIP for short) or surface mount. You can get Uno boards with either the DIP or *surface mount* (SMT) microcontroller. From an operational standpoint there's no difference between the two, except that with the DIP version you can yank the microcontroller chip out of its socket and replace it with another. The SMT chip cannot be readily unsoldered and replaced.
● The latest mega form-factor boards use an *Atmel ATmega2560*. These microcontrollers are more robust, supporting about two and a half times the external connections as boards that use the '328, and four times the memory.
● Some older Arduino reference design boards—such as the Duemilanove—and quite a few low-cost Arduino clones use an *ATmega168*. It's similar to the '328 found in the Uno, but has half the memory space. With less memory you're limited to smaller programs.

Sockets for External Connections
A key concept of the microcontrollers like the Arduino is that it's designed for *physical computing,* which means a computer meant to directly interact with the physical world around it. Microcontrollers achieve that through external connections. These are the thoroughfares that the controller uses to interact with other electronic devices.

To make it easier to experiment with different circuit designs, the basic Arduino—like the Uno—uses a series of 28 female *pin headers* that allow connection to other circuits. The headers are separated into three groups, as shown in Figure 1-5. These groups are

● Power
● Analog input
● Digital input and output

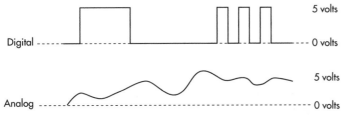

Figure 1-6 Digital versus analog voltages. These are the two types of signals the Arduino is designed to work with.

I'll save the power connections for the "Built-in Voltage Regulator" section later, and move right to the analog and digital inputs and outputs.

Of the 28 external connection pins, 20 are devoted to the input and output of signals—they're collectively referred to as input/output pins, or *I/O*. These pins are the connection points (call them nerve pathways if you'd like) between the Arduino brain and the rest of your robot. Through these I/O pins you connect motor circuits, sensors, and other electronics.

There are six analog input pins, labeled A0 to A5. They're engineered to accept a varying voltage (see Figure 1-6), typically 0 to 5 volts (but not always; see the section later on Arduinos that are made to work at a different voltage). Analog inputs are commonly used with sensory circuits, like photoelectric eyes.

The 14 digital input and output pins—labeled simply by number starting with 0—are located along the top of the basic Arduino board. They work with two voltage levels only, 0 or 5 volts. These voltage levels correspond to the digital states *off* and *on*, also commonly referred to as *LOW* and *HIGH* and *0* and *1*. They all refer to the same thing.

Volts	Synonyms Used in Digital Circuits
0	off, LOW, 0
1	on, HIGH, 1

Note that the 14 I/O pins, or *I/O lines,* can function as digital inputs *or* digital outputs. When acting as an input, the I/O senses when something connected to it is at 0 or 5 volts. A practical example of this is a switch. When open, its output might be 0 volts, but when closed (pressed), its output might be 5 volts. This is how a microcontroller can determine the current state of any kind of switch.

When acting as an output, the I/O provides 0 or 5 volts to some external device. A common application is to illuminate a light-emitting diode (LED). At 0 volts the LED is off, but given 5 volts it turns on.

These are simplifications of how digital I/O is used to interact with external components, but it'll suffice for now. In future chapters you'll learn specifically how to connect switches, LEDs, and a multitude of other components to the Arduino's I/O pins.

Other Arduino form factors have fewer or greater I/O lines, and not all expose the I/O lines as female pin connectors. Example: the Mega2650 has 54 digital input/outputs and 16 analog inputs. Extra pin headers provide the additional connection points.

The Pro Mini, being small and meant to be incorporated in a larger circuit, doesn't come with female header pins. You have the option of soldering wires directly to the I/O connection points on the board, or attaching your choice of male or female headers. When the headers are attached onto the underside of the board, you can plug the Pro Mini into a solderless breadboard; when attached to the top of the board, you can string *jumper wires* to whatever connection points you wish to use.

To provide maximum flexibility the analog inputs also serve double duty as ordinary digital inputs/outputs. In this way, the basic Arduino board supports 20 digital I/O lines, 6 of which can also function as analog inputs.

USB Connection

The Arduino is intended to be programmed using a personal computer. Software running on the computer communicates with the Arduino via a standard USB connection. This connection actually serves three important functions:

Program upload   As already noted, programs you write are uploaded to the Arduino development board via USB.

Arduino-to-PC communication   Seldom do microcontroller programs work perfectly the first time around. During the testing phase you often need a way to see what's going on inside the Arduino. This process, called *debugging,* is aided by having your Arduino transmit messages back to your PC, where you can verify things are working as they should (and if not working, why).

Temporary power   The USB connection provides a stable source of 5 volts to the Arduino, saving you from providing a separate power source during program development. Once your program has been perfected, the Arduino is untethered from its host computer. You need to power it from a set of batteries or other source.

Like many USB devices, the Arduino requires a *driver* that tells your computer how to communicate with it. This driver is included in the Arduino software, and complete instructions for installing this driver on various operating systems (Windows, Macintosh OS X, or Linux) are available on the main *www.arduino.cc* Web site.

Once the driver is installed you need merely to connect a suitable USB cable between the Arduino and your computer. The cable even provides the power to the board. The Arduino Uno uses a USB jack that accepts a standard *USB Type B* connector. This is a fairly fat thing so it's hard to miss. Your PC probably uses the larger *Type A* connector, so you need a Type A–to–Type B USB cable.

A few Arduino boards, and many USB adapters and cables, use USB *Mini* connectors. When using one of these boards you'll need a cable outfitted with the appropriate connectors on either end. So check first before purchasing a cable for use with the Arduino. Other Arduino boards use a USB Micro connector. USB Mini and Micro connectors look similar, but cables for one won't fit the other. Check the hardware guide for your board at *www.arduino.cc* for the details on which type of USB connector it uses so that you can be sure you're plugging in the right kind of cable.

Not all Arduino boards come with a USB connection. By leaving off the USB circuitry the board is less expensive, and likely smaller. In order to program the Arduino you need one of the following:

- *USB module or cable.* These have USB circuitry built into them, and plug directly into the Arduino board. As USB circuits add an average of $5 to the cost of the Arduino board, the idea here is that you buy one USB module or cable, and use it with any number of less expensive Arduino boards. Caveat: *The USB module or cable must have a compatible connector for attaching to your particular Arduino board.* You also need to be sure the USB module or cable is designed for the same voltage your Arduino uses, either 3.3 volts or 5 volts. See the section "Of Volts and Speed," later in this chapter.
- *Serial interface cable and connection adapter.* At its heart the Arduino communicates with your computer using good old-fashioned *serial* (one *bit* at a time) data signals. You can bypass the USB stuff entirely if your computer has one or more available 9-pin serial ports on it—possible with older Windows-based PCs and laptops, but virtually unheard of on the Macintosh.

Important! You still need a USB driver to use a USB module or cable. What's more, the USB driver that comes with the Arduino software will likely not work with a separate USB module, as it's not made for the same hardware. Check the site where you obtained the module or cable for the USB driver you're supposed to install.

By using the older-fashioned 9-pin serial port on your computer (if it's so equipped) you don't need to install a USB driver. However, since you're not using a USB connection you need to provide separate power (batteries, wall transformer) to the Arduino board.

Built-in Voltage Regulator

Electronic circuits need power to run, and that power most often is from a battery or wall transformer. The voltage provided by this power source depends on the battery or transformer, but 6 to 12 volts is not uncommon. Some kinds of electronic circuits can tolerate a wide range of voltages, and they don't care.

Not so with the Arduino. It runs on a strict diet of 5 volts.* Less and the microcontroller may malfunction; more and the Arduino circuitry may overheat and become permanently damaged. An on-board voltage regulator conforms the power from a battery or wall transformer.

The battery or wall transformer may be plugged into the 2.1mm power jack on the side of the Arduino board. (On those Arduinos not equipped with a power jack the incoming voltage is connected to terminals labeled Vin, RAW, or something similar.)

Typical operating voltage of the Arduino is 5 volts, which is supplied either by the USB cable when it's plugged into a USB port on your computer, or by a built-in voltage regulator when the board is powered externally. The regulator is intended to be powered by 7–12 volts DC (*direct current,* like that from a battery). For basic experimenting a 9-volt battery is ideal.

See Table 1-1 for a review of the power requirements when using a wall transformer, like the kind in Figure 1-7, or battery connection.

---

* Some models of the Arduino are meant for operation at 3.3 volts rather than 5 volts. These variations are discussed in "Of Volts and Speed," later in this chapter.

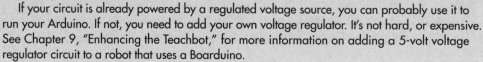

Table 1-1 Power Requirements for Most Types of Arduino Boards	
Minimum voltage	7
Maximum voltage	12
Voltage type	DC (direct current)
Polarity of power supply plug	Center positive
Plug type	2.1mm barrel

Avoid using a power supply of more than 12 volts, or it could cause the Arduino's voltage regulator to overheat. And be sure the output of the power source provides DC (direct current), and not AC (alternating current). Some wall transformers are designed to provide AC voltage rather than DC, so be sure to check. Otherwise, damage to your Arduino is likely.

Not all Arduino-like boards made by third-party manufacturers have on-board voltage regulation. The Adafruit Boarduino is a good example. This is not a bad thing, per se, but you need to be careful not to apply more than 5 volts to the board, or else damage to the microcontroller and other components could result.

If your circuit is already powered by a regulated voltage source, you can probably use it to run your Arduino. If not, you need to add your own voltage regulator. It's not hard, or expensive. See Chapter 9, "Enhancing the Teachbot," for more information on adding a 5-volt voltage regulator circuit to a robot that uses a Boarduino.

As you read earlier, when plugged into a USB port on your computer, the Arduino can derive its power from your computer. You don't need a separate battery or wall transformer supply in this case, though if you plug one into the 2.1mm jack, your Arduino will derive its power from it, rather than from your PC's USB port.

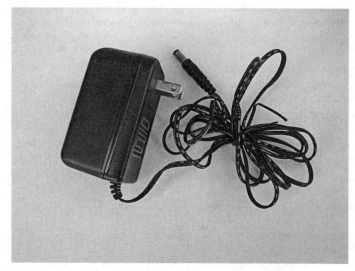

Figure 1-7   A common wall transformer, sometimes called a "wall wart," for providing external power to the Arduino. The 2.1mm barrel connector plugs into the jack on the side of the Arduino board.

As shown in Figure 1-8, there are eight pins dedicated to the Arduino's power connections on the basic Uno (and similar) boards. Five power pins are located along the bottom, and two are in the upper left, next to the digital input/outputs. They're labeled for easy identification, but let's cover them here:

- 5V is the voltage provided by the Arduino's on-board 5 volt regulator. You may connect other circuitry requiring 5 volts to this power pin, but be sure to observe current limits. See the accompanying warning for details.
- 3.3V is the voltage provided by the Arduino's secondary on-board 3.3 volt regulator. This regulator is primarily intended to power the USB circuitry, but its output is also available in case you need to provide power for some electronics that require 3.3 volts.
- Gnd (three pins total) provide a ground for your connected circuitry. Ground is also referred to as 0V (for 0 volts), common, earth, and a bunch of other names. All three ground pins on the basic Arduino board are connected together, so you can use any of them.
- Vin is the raw, or unregulated, voltage provided through the 2.1mm power jack.

The Arduino's on-board regulators can only supply a certain amount of *current* before overheating. Current, expressed in *amps* or *milliamps* (thousandths of an amp), is the measure of electrical charge flowing through a wire or other circuit. Higher currents demand larger wires, larger regulators, larger just about everything.

The Arduino 5 volt regulator can handle up to 800 milliamps (0.8 amp) of current; the 3.3 volt regulator, only 50 milliamps (0.050 amps). If you require regulated voltage at higher currents you'll need to use external voltage regulators.

Extra ground connection
AREF connection

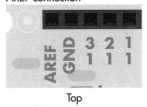

Top

Power jack

Left side

5V and 3.3V connections
Ground connections
RAW Vin connection

Bottom

Figure 1-8   Power connection pins on an Arduino Uno board. These are located along the side, bottom, and top of the basic Arduino board.

Note two other pins among the power connections on the Arduino. These are marked Reset and AREF. The *Reset* pin is connected with the reset button on the Arduino. It's there in case you want to use a reset pin located elsewhere.

AREF is used to provide a reference voltage for the analog inputs. Normally, the analog inputs expect a voltage range of 0 to 5 volts (assuming an Arduino that's operating at 5 volts). However, some kinds of sensors don't provide the full 0–5 volt range. By applying a lower voltage to the AREF pin you can effectively adjust the sensitivity of the analog inputs.

But take heed! When using the AREF pin you must be *super careful* to apply a voltage only within the correct range, *and* programmatically set up the Arduino to properly read the new reference voltage. If you do things in the wrong order, or apply an incorrect voltage to AREF, you can permanently damage the ATmega microcontroller on your Arduino board!

None of the projects in this book require using the AREF pin, but I include this caution in case you want to experiment on your own.

AVAILABLE MEMORY

The ATmega microcontroller on the Arduino development board contains its own memory. This memory is used for holding the programs you develop, and for data access when a program is running. The amount of available memory varies, depending on the version of the ATmega chip that's used. For discussion purchases, what follows is for the ATmega328P, used in the Arduino Uno basic board.

Flash Memory

Flash is where the programs you write and store on the Arduino are stored. The ATmega328 provides 32KB (yes, that's *kilobytes,* not megabytes or gigabytes) of program storage. That may not seem like much, but in the world of microcontrollers that's usually plenty to get the job done. Of the 32KB of space, half a kilobyte is used by a special program, called a *bootloader,* that permanently resides in the flash memory. More about the bootloader program in "A Closer Look at Arduino Software," later.

Static RAM (SRAM)

The Arduino needs a place to store temporary data when a program runs. That's the job of the 2K (2 kilobytes) of static *random access memory* (RAM). Like the flash memory, 2K may seem wholly insufficient, but it's adequate for most jobs. When it's not, you can add external static RAM chips to expand the amount of memory available. (None of the projects in this book require extra RAM.) Data in RAM is lost when power to the Arduino is removed.

EEPROM

Additional data can be saved in the 1KB of EEPROM—which stands for *electrically erasable programmable read-only memory.* Unlike RAM, this data bank retains its memory even when the Arduino is unplugged from its power. EEPROM can be used for things like storing values for mathematical formulas, or to keep track of sensor readings taken over weeks, months, even years.

The three types of memory differ in whether data is retained when power to the Arduino is removed. This is called *data volatility,* and is summarized here.

Memory Type	Capacity*	Data Volatility
Flash	32KB	Data is *retained* after power is removed and/or the Arduino is reset.
Static RAM	2KB	Data is *lost* when power is removed or when the Arduino is reset.
EEPROM	1KB	Data is *retained* after power is removed and/or the Arduino is reset.

* Memory capacity is for the ATmega328P microcontroller chip. Other versions of ATmega microcontrollers have different amounts of memory. Consult the Hardware section on the main Arduino Web site for more information on memory provided in any of the reference designs. Manufacturers of custom Arduino-based boards will provide similar information.

Of Volts and Speed

Most flavors of Arduino operate at 5 volts and run at a processor speed of 16 megahertz. Just to complicate things, this isn't true of all Arduino boards, even the official reference designs. Here's what you need to know.

ARDUINO POWER: 3.3 OR 5 VOLTS

While 5 volts is a common supply voltage for modern electronic circuits, it's not the only game in town. Gaining in popularity are circuits and modules that are intended to run at 3.3 volts (some are made to work at even lower voltages, but these are not of concern here).

Certain Arduino boards are available in either 5 volt or 3.3 volt versions—for simplicity, I'll refer to these as 5V and 3.3V, and save a few letters of the alphabet. The operating voltage of the Arduino affects what external components it can be connected to. You'd want a 3.3V Arduino if most or all of the circuitry you're connected to requires 3.3 volts. Using a 5V Arduino to connect to circuits expecting 3.3 volts can cause damage to those circuits, so this is something you want to be mindful about.

The vast majority of Arduino reference design boards are made for 5 volt operation. Those that are intended for 3.3 volts include the Fio (made for wireless communication) and the 3.3V version of the Pro and Mini Pro. Some Arduino boards are made to accept a range of voltages: the Lilypad works with 2.2 to 5.5 volts, for example.

The more common (at least for now) scenario is to use a 5V Arduino and translate—or step down—the voltage when connecting to 3.3 volt devices. There are a number of ways this can be accomplished, including simply putting a resistor between the Arduino and its lower-volted companion. Several of the techniques for interfacing a 5V Arduino with 3.3 volt devices are demonstrated throughout this book.

ARDUINO OPERATING SPEEDS

All microcontrollers, Arduino included, use the pulses of a system clock to keep pace with their internal processing chores. This clock runs very fast in human terms; the most common clock speed of the Arduinos is 16 *megahertz*, or 16 million cycles per second (megahertz is commonly shortened to *MHz*). As a point of reference, the first versions of the IBM PC ran at 4 MHz; the Apollo guidance computer used to land men on the moon operated at just 2 MHz. Your pocket calculator now runs faster than that!

The ATmega chip on the Arduino uses each clock tick to process a programming instruction, or to pass a piece of data from one place to another. The architecture of the ATmega chip allows it to complete most programming instructions in a single clock tick (called *cycle*), meaning it can compute at the rate of up to 16 million instructions per second.

Quite often those Arduino boards that operate at a lower voltage also run at half speed, or 8 MHz. The same is true of most any Arduino board designed to consume low power; the faster the chip is operated, the more power it consumes.

Arduino's system clock is set by a crystal, encased in a tiny metal can—you can see it near the USB jack on the Uno board in Figure 1-5. The crystal vibrates at a very specific frequency, giving the clock a high degree of accuracy. Different clock speeds are attained by using crystals that run at different frequencies.

The ATmega328P—which, you'll recall, is used in most of the current Arduino boards—can also operate without a crystal, using a built-in resonator that provides the clock pulses. This resonator operates at up to 8 MHz. You'll often see simplified Arduino boards that run at 8 MHz. They lack a crystal, and instead rely on the ATmega's built-in resonator.

While 8 MHz is still plenty fast for many applications, the resonator is not as accurate as a crystal. Select a crystal-based Arduino if you absolutely need to accurately time such things as how long it takes for a sound wave to bounce off an object.

The ATmega328P chip is engineered to run at up to 20 MHz, a speed increase of 25 percent over the typical 16 MHz. This higher speed is not commonly used, though some makers of Arduino-compatible products offer 20 MHz speed as an option.

Yes, a 20 MHz Arduino runs faster, and therefore can crunch more instructions each second. But for most users, the added complexities of dealing with timing issues caused by the faster speed are not worth the speed benefits. You should only attempt to use an "overclocked" Arduino if you are comfortable modifying the Arduino software, in order to adjust for speed differences.

Look at All the Blinky Lights

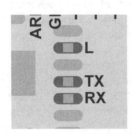

Light-emitting diodes are provided as indicators on the Arduino for testing and verification. A small LED shows power; two other LEDs show serial transmit and receive activity, and should flash when the board is being programmed from your computer.

A fourth LED is connected in parallel with digital I/O line 13, and serves as a simple way to test the Arduino and make sure it is working properly. This LED is often used in first-time sketches as a way to check basic programming functionality of the Arduino. Such a sketch, and others, is provided in Chapter 2, "Arduino Up and Running."

Older Versions of the Arduino Boards

The Uno is the current exemplar of the basic board reference design. There were several similar boards before the Uno hit the scene. They all look similar to the Uno of today, but they used different components and had different features.

Here's a quick recap of the more popular boards that are no longer made, just so you're kept well informed:

Arduino Serial   The birth of the Arduino. Connected to a PC using a serial port.

Arduino USB   Like its predecessor, but connecting to the PC using the (now) more common USB port.

Arduino Extreme   Additional features by incorporating tinier surface mount components.

Arduino NG   The NG stands for Nuova Generazione, which in Italian means *New Generation*. Folks started taking real notice of the Arduino with this version.

Arduino Diecimila   Programming the Arduino gets easier with the Diecimila, thanks to a new feature allowing it to be reset from the computer when uploading a new program (Diecimila is *10,000* in Italian; as in "10,000 Arduinos made to date").

Arduino Duemilanove   Further refinements, such as auto-selection of the power source—in previous USB versions you needed to set a switch between power-from-USB and power-from-power-jack. For the curious, Duemilanove is for *2009*, the year the board was introduced.

Arduino Stamp   Similar in appearance to the venerable Parallax BASIC Stamp 2 product, everything on a small circuit that's the same size as a wide 24-pin integrated circuit. The current Arduino Pro, while not called a "Stamp," or suggested as a BASIC Stamp alternative, shares the same pinout as the BASIC Stamp 2.

Arduino Mega   Very similar to the modern-day Arduino Mega2560, but used an ATmega1280 chip instead.

A Closer Look at Arduino Software

If you've ever used a microcontroller you know the process of programming it involves three steps: write the program, compile the program, and run the program. These steps are shown in Figure 1-9.

The Arduino is no different, except that it refers to its programs as *sketches*. Sketches are written in a programming language very similar to C, one of the world's most widely used computer languages.

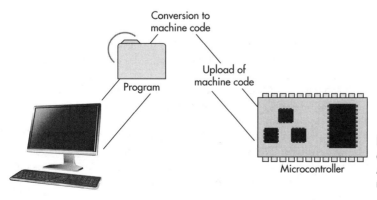

Conversion to machine code

Program

Upload of machine code

Microcontroller

Figure 1-9   The three phases of software development on the Arduino—or for that matter most any microcontroller

In fact, the programming language used with the Arduino is functionally like C, but with several simplifications to make it easier for newcomers to master the language. It also supports the notion of *classes* and *objects,* features found in the higher-order language C++—pronounced *see-plus-plus.* Both classes and objects are detailed in later chapters, including Chapter 4, "Programming the Teachbot: Making It Move."

If you've ever looked at a C/C++ program and felt your eyes glazing over because of the obtuse appearance of the code, you needn't worry about that with the typical Arduino sketch. The Arduino is designed for beginners in mind, but still provides power and flexibility for more advanced users. The Arduino programming language is inherently expandable. Like an iceberg, only a small percentage of the Arduino's programming power is immediately visible.

Looking more closely, the three steps of writing and uploading Arduino sketches are:

1. *Develop* your sketch on your PC. The Arduino comes with a Java-based integrated development environment—*IDE* for short—that includes a fully featured text editor. (In order to use the IDE you need to install the Java framework on your PC if you don't already have it. If you've used your computer any length of time you probably already have Java on it.) The IDE supports syntax highlighting and coloring—that is, different parts of code are shown in different colors. Figure 1-10 shows how the IDE looks with a sketch contained in it.
2. Once written, sketches must be *compiled,* which in Arduinoland is referred to as *verifying.* During the compile/verify phase any errors are flagged and noted at the bottom of the Arduino editor window. The compiling process includes building the sketch from component files. The Arduino does a remarkable job of simplifying and streamlining the compiling (okay, verifying) process, which historically has been one of the biggest sticking points in microcontroller programming.
3. The compiled program is then *uploaded* to the Arduino via a USB cable. The upload process is automatic. This is thanks to a *bootload* program that is permanently stored in the flash memory of the Arduino. This program detects whenever a new sketch is arriving from your PC. The bootloader performs the necessary steps of first erasing the old sketch in memory, if present, then accepting the new one. Once uploaded, the sketch starts automatically within the Arduino.

ABOUT IDE VERSIONS

The Arduino IDE and the standard programming statements and libraries often undergo changes with each new version. Versions are indicated by number, such as 1.0. Each time the IDE is updated its number is incremented.

If you already have an installed version of the IDE and it's old, you'll want to fetch the newest version. You can keep multiple versions of the Arduino IDE on your computer, and even switch between them as needed—though that should seldom be required.

Unless otherwise noted, the projects in this book require version 1.0 or later of the Arduino IDE. Unless you tell it not to, the Arduino IDE is set to check for the availability of updates. If you open the IDE and it tells you a new update is ready, download it and install, and be sure to take a look at the readme file included with it for the latest changes.

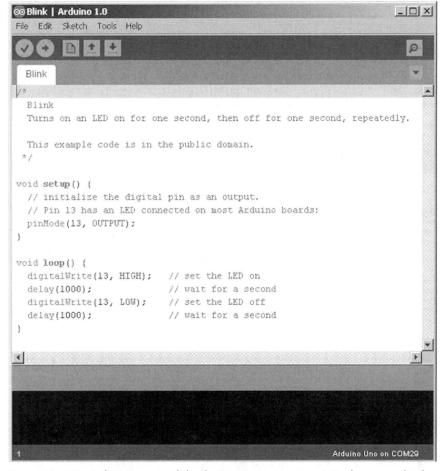

Figure 1-10   Arduino integrated development environment, or IDE, showing a *sketch* ready for upload to the board

 If you have a pre-release version—these use a four-digit number such as 0022—you'll want to update it before trying the sketches from this book. If you'd like, you can keep the old version and install the new one alongside it. This can be helpful if you have some existing sketches that rely on the older Arduino IDE software.

USB DRIVER

The necessary USB drivers to connect your "official" Arduino board with a host PC are provided with the Arduino software. In many cases, and depending on the operating system of your computer, installation of the drivers is not completely automatic. But the steps are straightforward, and the Arduino support pages provide many walk-through examples.

Ready Expansion via Breadboards and Shields

The Arduino is a good example of keeping things simple. Its no-frills design helps drive down costs, and makes the Arduino a universal development board adaptable to just about anything. The basic Arduino board lacks connectors to directly attach to motors, sensors, or other devices. And while there are more expensive specialty versions of the Arduino intended for robotics applications, many users don't opt for them.

Instead, they attach these and other external components using a variety of other means, such as a *solderless breadboard,* shown in Figure 1-11. If you've experimented with electronics no doubt you've used a solderless breadboard to build and test circuits. Components and wires stick into contact points in the breadboard to create fully functional circuitry.

I rely heavily on solderless breadboards throughout this book, for a number of reasons. First, solderless breadboards allow a great deal of experimentation. You can always pull out one component and plug in another. Soldering is kept to a minimum, so changes are easier. Breadboards also provide more latitude to deal with mistakes. When you hardwire things into a circuit it's more difficult to undo any errors you make.

But perhaps one of the main reasons I like solderless breadboards is that it's easier to reuse components. You can easily share parts between your robots.

Are there downsides to solderless breadboards? Yes, and many—like wires that can fall out or come loose—are covered in more detail in Chapter 8, "Making Things: Electronic." But if you're careful, these limitations are easy to work around, even avoid completely.

Figure 1-11   One of many types of expansion shields designed for use with the Arduino Uno and similar boards. This one combines a stackable shield with a mini solderless breadboard. You connect components on the breadboard and contact points on the shield using short jumper wires. *Photo courtesy Adafruit Industries.*

Another popular method for expanding the Arduino is through separate small circuit boards called *shields.* These stick directly on top of the board. Pins on the underside of the shield insert directly into the Arduino's I/O headers. Two popular expansion shields are the solderless breadboard and the proto shield; both provide prototyping areas for expanding your circuit designs.

What's more, you can often stack multiple shields right on top of one another. You might combine a shield for controlling motors with another for operating a series of LEDs, plus a third that has its own backlit LCD display panel built in.

Shields may be combined as long as there aren't any conflicts in the I/O pins used by each one. Except in rare circumstances, if an I/O pin is used by one shield, it can't also be used by another. (See Chapter 8 for suggestions and alternatives for when shields collide.)

Chapter 2

Arduino Up and Running

So you read in Chapter 1 that the Arduino is a development platform with a programmable microcontroller at its heart. You write programs for the Arduino using your PC. When you're done, you send those programs to the Arduino, where they can operate independently of the PC. This is what allows you to use an Arduino in a mobile robot, without keeping it tethered to your computer.

Simple enough to say all this, but a bit more involved to actually do it. That's where this chapter comes in. If you're brand new to all things Arduino and electronics, or you need a refresher course, this chapter will get you started in the right direction. It contains a quick guide to downloading and installing the Arduino software, plus four simple yet fun first-time examples that demonstrate basic Arduino tinkering. All four examples are designed to demonstrate functionality important in robotics.

FYI This chapter is a fast-start guide to getting started with programming the Arduino. It's intended for beginners, and it introduces and/or repeats basic microcontroller and programming terminology, in case these concepts are new to you.

By all means, if you've already mastered this material, feel free to skip ahead. Starting with the next chapter you'll learn how to construct the Teachbot, an expandable desktop robot that uses an Arduino as its processing brain. The Teachbot is intended to be a *platform* for trying out different robotic techniques.

Parts You Need...

Besides your computer or laptop, you'll need what you see in Figure 2-1 in order to complete the tasks in this chapter.

- A reasonably up-to-date Arduino development board. For the sake of following along with the tutorial in this chapter I'll assume you're using an Arduino Uno, or another basic Arduino board very close to this model. (Note: Don't use an Arduino Leonardo. Though it looks like an Uno board, it has several under-the-hood differences. This caution applies to all the projects in this book. Uno yes, Leonardo no.)

- A USB cable for connecting the Arduino to your computer. The Uno uses a Type B jack; most PCs use a Type A jack. Therefore, you need a cable with a Type A plug on one end and a Type B plug on the other. For ease of use the cable should at least three feet long.
- A solderless breadboard. A mini breadboard with 170 contact points is sufficient. (See Chapter 8, "Making Things: Electronic," if you're not familiar with solderless breadboards or how to use one.)
- A collection of jumper wires for the solderless breadboard. You can make your own 22 gauge solid conductor wiring, or purchase a small set where the wires are pre-cut and pre-stripped. (Again, see Chapter 8 for more information on this topic.)

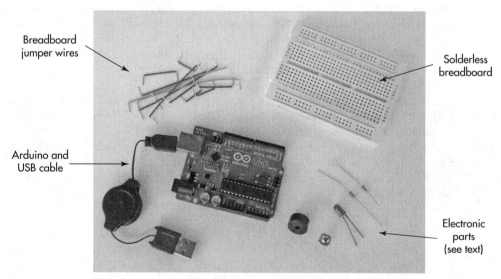

Figure 2-1   The parts you need to complete the example projects in this chapter. Note the style of the solderless breadboard. It's the recommended type for the Teachbot learning platform, detailed in the following four chapters.

- These assorted electronics parts:

1	Standard size red light-emitting diode (example: Radio Shack #276-330)
1	470 Ω resistor (Radio Shack #271-1317)
1	10 kΩ resistor (Radio Shack #271-1335)
1	Miniature piezoelectric speaker (example: Parallax #900-00001)
1	Tactile ("tact") switch (Parallax #400-00002)

Consult Appendix B, "Parts Connection," for a list of several online sources providing these and other electronic components.

 You can get these parts in a handy all-in-one kit from several online retailers. See Appendix A, "ARB Support Site," for more information.

Arduino Quickstart

Recall from Chapter 1 that the Arduino is a full development platform consisting of both hardware—the board itself—and software. Download and install the software on your computer *before* connecting the Arduino board to your computer.

Depending on the operating system of your PC, the installation process is not difficult, but there are variations in the exact steps. Following is a general guide so you know what's involved.

 The Getting Started pages on the main Arduino Web site do an excellent job of providing step-by-step help for installing the Arduino software in several versions of Windows, as well as Mac OS X and Linux. I see no point in spending valuable book pages repeating what's already been written and widely available.

Visit *www.arduino.cc*, and click on the *Getting Started* tab. Click on the link for your operating system. You may even wish to print out the pages so you can have them beside you as you follow the steps.

STEP 1—DOWNLOAD THE ARDUINO SOFTWARE

Use your browser to download the Arduino software package by going to the Arduino Web site (*www.arduino.cc*), then clicking the *Download* tab. Find the link for the latest version for your computer's operating system.

 The Arduino software is available in ready-to-go executable format and source code format. You want the former. *Executable* files are ready-to-go programs; you just double-click on the file and the program runs. *Source code* is the actual programming that goes into making the executable files. It's provided for advanced users, who may wish to modify the software for their needs.

Some operating systems offer multiple versions, depending on type; for example, either 32-bit or 64-bit versions. Be sure to get the version for your operating system. A 64-bit version of the Arduino software will not run on a 32-bit operating system.

STEP 2—UNPACK THE SOFTWARE COMPONENTS

The Arduino software is distributed in an archive file—*zip* for Windows, *zip* or *dmg* for Mac, and *tgz* for Linux. These are all compressed files, combined into a single archive. They require unpacking before you can use what's inside.

Very likely your operating system already has a program for unpacking the archive, so just double-click on the archive file to start the process. If your operating system does not have suitable software, you can install a program for this task. A popular free alternative for Windows is 7-zip (*7-zip.org*).

For ease of use, I recommend putting the Arduino software in a folder off the main directory of your main hard disk drive (you may need administrator/owner rights to do this). Follow these steps:

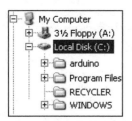

1. After downloading, copy the archive to the *root* of your computer's main hard drive (on Windows machines this is the C: drive); that is, outside any directory folders.

2. Unpack the archive into a folder off the root. A new folder under the root will appear. This folder will contain all the Arduino software files.

3. When done, delete the archive file. You don't need it anymore, but even if you do, you always know where you can get a copy.

STEP 3—INSTALL THE USB DRIVER

The Arduino uses USB as the communications link between it and your computer. Your PC doesn't know how to talk to the Arduino without an interpreter, and the USB *driver* included with the Arduino software distribution does just that.

Here's where the exact steps differ depending on your operating system, and even between versions of the operating system. With Windows XP, for example:

1. Plug in the Arduino using the USB cable.

2. The computer recognizes that a new type of hardware has been connected, and it'll display a step-by-step wizard to help guide you through the process.

3. When prompted, ask to specify the location of the drivers. You want the *drivers* folder under the main Arduino directory.

4. Click *Next* to complete the wizard. If the proper driver was found, Windows will indicate that the new hardware is ready for use.

The Getting Started pages on the Arduino Web site provide a different set of steps from those noted here. Though the steps on the Arduino Web site are more lengthy, they're more likely to be consistent across different versions of Windows.

There's also a guide with screenshots that is specific to Windows XP. You may find this page from the main *arduino.cc/en/Guide/Windows* page.

Keep these points in mind when installing USB drivers for the Arduino:

- Be aware that the main Getting Started section of the Arduino.cc pages assumes you're using an Arduino Uno or similar basic board. If you're using another version of the Arduino, be sure to check out its corresponding page on the site.

- Updates to the Arduino software can change the way you install the USB driver for any given operating system. That's why you should always refer to the main Arduino Web site for the latest essentials.
- Different versions of the Arduino hardware may require different USB drivers. Be sure to download and install the correct USB driver for the hardware you are using.
- If you are using an Arduino-compatible board, be sure to get the USB drivers for it from the manufacturer or retailer.
- Once you've successfully downloaded and installed the USB driver for a given Arduino board, you don't need to do it again, even if you later update the version of the Arduino IDE (though some exceptions can apply).
- In many cases USB drivers are *port-dependent.* Which means if you plug the Arduino into a different USB port on your PC, you may need to repeat the driver installation process. Since this is a hassle, get into the habit of always using the same physical USB port on your computer.

STEP 4—VERIFY YOUR ARDUINO IS WORKING

The Arduino gets its power from the USB connection when connected to a PC via USB. Without doing anything else you can use this opportunity to ensure that your Arduino board is working properly.

When you plug in your new Arduino look for the two indicator lights, as shown in Figure 2-2.

Power on indicator   At the very least when the Arduino is plugged into your PC via USB the power light-emitting diode (LED) should be lit. If it's not, that could indicate a faulty Arduino, bad cable, or busted USB port. See Appendix C, "Troubleshooting: When Things Go *Worng*," for ideas on how to fix these and other problems.

Pin 13 LED   If yours is a brand new, right-out-of-the-box "official" Arduino, it will come with a pre-programmed "LED blink" program. The LED should flash on and off once a second. (Note: It's called the Pin 13 LED because that's the input/output pin connected to the built-in LED.)

STEP 5—START THE ARDUINO IDE

Once installation of the software and driver is complete you're ready to download a new program into your Arduino. Start the Arduino integrated development environment (IDE) by double-clicking on the Arduino program icon. After loading, the IDE window appears, as shown in Figure 2-3. Main points of interest are noted.

Pin 13
LED

Power
on

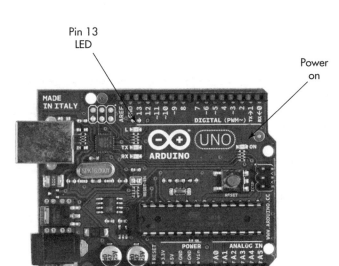

Figure 2-2 Important indicator lights on the Arduino Uno (and similar) development boards. They confirm proper operation of the board.

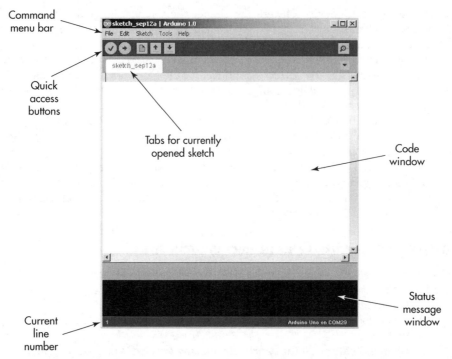

Command menu bar

Quick access buttons

Tabs for currently opened sketch

Code window

Status message window

Current line number

Figure 2-3 The Arduino integrated development environment (IDE) is simplicity to its core. The code window is where you write and edit your Arduino sketches.

If not already connected, plug in the USB cable between the Arduino and PC.

In order to use the Arduino IDE your computer may need a recent installation of the Java framework. You may receive an error if your computer doesn't have Java, or the version of Java installed on it is old.

As required, download the most recent version from the main Java Web site at *www.java.com*. It's free, but be careful about any "bonus" software that the Java installer may add. Typical are search engine toolbars. You will have the option of not including this additional software when Java is installed. You must specifically uncheck the box to prevent the software from being added to your computer.

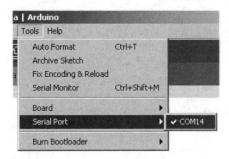

STEP 6—SPECIFY YOUR ARDUINO BOARD AND COMMUNICATIONS PORT

The Arduino IDE requires virtually no setup prior to actually using it to make programs for the Arduino, but there are two important steps you have to do first:

Specify your Arduino board Recall there are numerous variations of the Arduino board, and the IDE needs to know which board you're using. Select your board under *Tools | Board*. For example, when using the Uno you select *Arduino Uno* in the board menu.

Specify the communications port connected to your Arduino You may have more than one active communications port, and the IDE needs to know which one is attached to the Arduino. Select the port under *Tools | Serial Port.* The port currently selected by the IDE has a checkmark beside it.

If you're not sure which port to use, try unplugging the Arduino from the PC, then note which port disappears from the Serial Port list. The one that goes away is the one associated with your Arduino. Plug the board back in, and select that port.

What? The Serial Port menu option is dimmed and can't be selected? That means the Arduino IDE does not recognize any serial port is currently active. Double-check that your Arduino is plugged into a working USB port and that its indicator light is on. See Appendix C, "Troubleshooting: When Things Go *Worng*," for additional help.

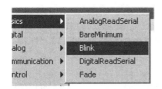

STEP 7—LOAD A TEST SKETCH

The Arduino IDE refers to programs as *sketches*. You can quickly test your new setup by loading a simple sketch into the Arduino.

1. In the Arduino IDE, choose *File | Examples | Basics | Blink.* In a new window the good ol' blink-the-LED sketch appears. It's a good test to ensure everything—including your Arduino—is working.

2. Click the *Verify* button. Verifying is what the Arduino calls compiling a program. Status messages are displayed at the bottom of the IDE window as the sketch compiles.
3. After successful compiling (verifying), you may transfer the sketch to the Arduino by clicking the *Upload* button. As with compiling, the window at the bottom of the IDE displays status messages.

FYI Things didn't work the way you expected? See Appendix C, "Troubleshooting: When Things Go *Worng*," for a list of common errors and how to correct them.

STEP 8—MODIFY THE TEST SKETCH

The Blink sketch is the same one pre-loaded on most new Arduino boards. So if you're working with a brand new Arduino board, you may not notice any difference if you compile and upload the same sketch.

As a simple proof-of-concept, make a quick change to the Blink sketch by looking for, and changing, both program lines that say

```
delay(1000);
```

Change the number 1000 to 2000:

```
delay(2000);
```

Make no other changes. Recompile and reupload the sketch. Now the LED should blink more slowly, once every two seconds.

You cannot save over the example sketches. If you try, the Arduino IDE will tell you the sketch is read-only. You'll be given an opportunity to make a new sketch of the example. Unless you tell the IDE otherwise, your sketches are stored in a new Arduino directory under your user documents folder.

You can open and save sketches anywhere on your computer that you have access rights to, but I like to keep things consolidated, and use the Arduino documents folder for pretty much everything. Keeping things in one place also helps you make quick backups. To make a backup just copy the contents of the Arduino documents folder to a thumb drive, external hard drive, or recordable DVD.

Looking Again at the Arduino

Though this is covered in Chapter 1, I'll repeat the more salient points regarding the physical layout of Arduino. It's important stuff because your programs (er, sketches) interact with the Arduino hardware. The main points of interest are:

Input/output (I/O) pins   The Arduino provides a total of 20 I/O pins for connecting with the outside world. On a standard board like the Uno, they're labeled A0 to A5 and 1 through 13. Recall from Chapter 1 pins A0 through A5 accept analog (variable voltage) input; all the other pins are suitable for use with digital off/on signals. In the sketches you write you'll reference these pins by their numeric nomenclature: *A1* for the pin marked A1, for example, or *13* for the pin marked 13.

When describing how a sketch works I'll refer to the digital pins as *D#*, where the # is a number like *D6* or *D13*. This is to differentiate these pins from the analog I/O. Most Arduino boards don't use "D" to denote the digital pins, but for the sake of consistency I'll often include it when discussing the functionality of a sketch.

Power pins   Some things you connect to need a source of power, and as long as the power demands aren't too stringent, operating juice can come from the Arduino. As noted in Figure 2-4, there are pins for regulated 5 volts, regulated 3.3 volts, ground (0 volts connection), and Vin, which is the raw voltage used to power the Arduino. This raw voltage comes either from the USB port—which is typically 5 volts—or from a battery or power pack connected to the external power jack.

You can also provide regulated voltage to the power pins, but you must be vewy, vewy caweful when doing it. Applying more than 5.5 volts to the 5V pin can cause permanent damage to the Arduino. Do not apply more than 12 volts to the Vin pin.

Reset button   Press Reset when you want to restart the sketch in the Arduino. This button is also referred to as a *soft reset,* or *restart.* You can also reset the Arduino by disconnecting, then reconnecting, its power.

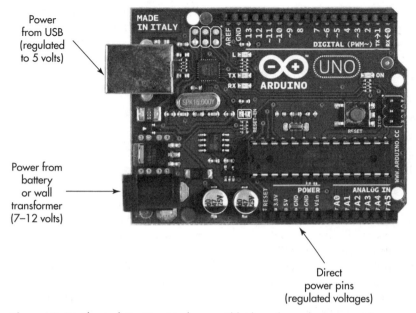

Power from USB (regulated to 5 volts)

Power from battery or wall transformer (7–12 volts)

Direct power pins (regulated voltages)

Figure 2-4   The Arduino Uno (and compatible) boards can be powered from a number of different sources, including from the USB connection to your computer, from an external battery or power supply source, and from direct connection to its power pins. The latter require carefully regulated voltage.

Whenever you upload a sketch the Arduino IDE externally manipulates the Arduino reset, allowing for the new sketch to be loaded into the Arduino's memory. On some versions of Arduino boards (especially the "non-official" ones), you may have to momentarily press the Reset button in order for the sketch to finish uploading. This is due to the architecture of the board, or the way the board connects to your computer.

In this book I assume you're using a standard reference design board like the Arduino Uno, where this "reset button dance" is not required. If it is needed for your board be sure to consult the documentation that came with the product. It'll provide the simple steps you need to coordinate the manual Reset button depress with sketch uploading.

Getting Started with Microcontroller Programming

As you by now know, microcontrollers depend on a host computer for developing and compiling programs. The software used on the host computer is the IDE—integrated development environment. We've already seen the one used with the Arduino. This IDE is based on an open-source project called *Processing* (*www.processing.org*), which is described by its creators as a "programming language and environment for people who want to program images, animation, and interactions."

The Arduino programming language leverages an open-source project known as *Wiring* (*wiring.org.co*). The Arduino language is based on good old-fashioned C, one of the world's most popular and commonly used programming languages. If you are unfamiliar with C don't worry; it's not hard to learn—you'll be doing it throughout this book—and the Arduino IDE provides feedback when you make mistakes in your programs.

TERMS AND CONCEPTS: IT'S ALL ABOUT DIGITS

Before diving into your first Arduino sketch, here's a quick tour of some programming terms and concepts that are handy to know. These and others are described in more detail throughout the book.

Digital defines electronic circuitry that has two finite conditions, usually indicated as *off* and *on*. Digital circuits can be either inputs (they read data) or outputs (they provide data). Because there are only two states in the digital world, all digital data is *binary*—capable of just two conditions.

Digital logic is a generic term that describes how a circuit or program reacts to changes in its inputs or outputs. Because there are only two possible outcomes for a digital input or output, digital circuits follow a well-defined *logic*. So, digital circuits are often referred to as "logic circuits," because of the Mr. Spock–like logical role they play.

Digital logic states refer to the two conditions possible for a digital input or output: off or on. Other terms for the same thing include:

off	on
LOW	HIGH
0	1

As with most books about programming and electronics, in most cases I'll use the LOW/HIGH and 0/1 nomenclature for digital inputs and outputs. Electrically, LOW (0) is usually 0 volts (also called ground or GND); HIGH (1) is usually 5 volts.

HOW HIGH IS HIGH?

The exact voltage for HIGH (or *on* or *1*) depends on the digital circuit. Most circuits are designed to be powered from 5V, so the HIGH state is 5 volts. But some are designed for operation at 3.3V, making their HIGH state equal to 3.3 volts.

Adding to the confusion is that for the Arduino, and many other circuits that rely on digital signals, the breakpoint between LOW and HIGH is not exactly 5V or 3.3V. More often, anything over a certain percentage of voltage—say halfway, or 2.5 volts—is considered HIGH.

You can sometimes use this factoid to your advantage, and in fact, we will in a few of the projects in the succeeding chapters. But for the sake of discussion, in this book (as with most), unless noted otherwise LOW is considered 0 volts and HIGH is considered the operating voltage of the circuit, usually 5V or 3.3V.

There's one other state for digital inputs and outputs: *floating*. This state is neither LOW nor HIGH. Digital I/O that floats is also called *hi-z*, or *high impedance*. There are times when you want this floating state. For example, a digital input is always floating. The digital state of an input is determined not by itself, but by some other circuit connected to it.

DATA BITS, BYTES, WORDS, AND MORE

This part is rudimentary to most readers, but I'll cover it for the sake of completeness: like all computer-based systems, the Arduino deals with data. There are different kinds of data, most commonly expressed by the number of bits it contains.

A bit is the smallest unit of data. It's composed of a single binary digit, which remember can be off or on (LOW or HIGH, 0 or 1). Think of it like a switch: in one position the switch is off, and in the other position the switch is on. A binary digit cannot have any other value. In the Arduino, the most common use of the bit is when working with digital I/O pins.

Bytes, words, longs, and other forms of data are composed of different amounts of binary bits (see Figure 2-5). The more bits, the more information that the data can store. A *byte* is the smallest collection of bits commonly used with the Arduino. One byte contains 8 individual bits. As each bit has just two possible values, you can mix-and-match the eight bits to make 256 possible combinations. Remembering that 0 is a valid value in computers, this enables the byte to hold any value between 0 and 255.

The Arduino uses over a dozen common data types, but the four most used are:

Data Type	Number of Bits	Largest Number It Can Hold
byte	8	255
int	16	65,536
long	32	4,294,967,295
boolean	8	true or false*

* Boolean data is meant to portray either of two conditions, true or false. A bit could do that too, but in the Arduino (and most computer programming languages), boolean data is actually stored as an 8-bit value.

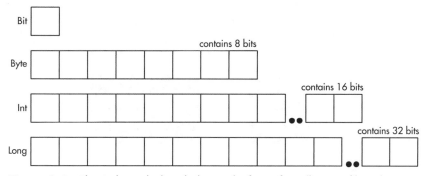

Figure 2-5   The Arduino deals with data in the form of a collection of bits; the most common data types are the single bit, byte (8 bits), integer (16 bits), and long (32 bits).

There are signed and unsigned versions of many data types. What's the difference? *Signed* numbers can be both positive and negative values, such as –15 or –40. *Unsigned* numbers can only be positive. An unsigned number cannot hold the value –40.

NERD STUFF: COMMANDS, FUNCTIONS, STATEMENTS, DATA TYPES, OPERATORS, AND VARIABLE NAMES

In computer programming, *keywords* refer to identifiers or instructions that have special meaning. They are quite literally the "key words" you use to construct your programs.

The most common keywords are:

- *Commands* are basic instructions you provide to tell the sketch what to do. A typical instruction is an *if conditional test,* which checks whether a condition is true or false.
- *Built-in functions* are like commands, but they're unique to the Arduino—or at least, to microcontrollers like the Arduino. Commands like *if* are more-or-less universal in C and many other programming languages. But functions like *digitalRead*—which you'll learn about later and in subsequent chapters—are specific to the Arduino.
- A *statement* (at least as used in this book) is composed of one or more commands. A statement may also contain one or more operators (see later) that instruct the Arduino what to do with those commands. The *if* keyword is always used in a full statement; the keyword *if* all by itself in a program is meaningless.
- *Data types* indicate the type and size of the data you're manipulating in your sketches. There are different data types to accommodate different kinds of numbers and other data. Since memory in a microcontroller is limited, you always want to pick the smallest data type that will hold the data. Otherwise, you'll unnecessarily use up valuable memory space.
- *Operators* tell the program how to manipulate and respond to data. The most common are *comparison operators,* for comparing if data is equal, not equal, less than, and greater than.
- *User-defined variables* are temporary holders for data being manipulated by your sketch. They're called user-defined because you make up their names. By storing data in a variable (see Figure 2-6) you can reuse the same data many times, simply by referring to the variable by name. With most types of variables, you can change their content at any point in the sketch. That's what makes them so powerful.

Other forms of keywords include *utilities* and *variable scope qualifiers.* I discuss these and others throughout this book as they relate to actual robotics projects, but if you want a more scholarly exposure to them, check out the Reference pages at the main Arduino Web site.

Variable named "X"   Variable "X" contains 10

Figure 2-6 Variables are temporary storage spaces for data. They are referenced by name.

SYNTAX AND STRUCTURE

All languages follow a *syntax,* which is how words are strung together to make understandable speech. You can understand the words when they're in their proper order. It's what differentiates "where no one has gone before" from "has before where gone one no." One sorta makes sense, the other not at all (unless you're Yoda, but that's a different movie franchise, so it doesn't count).

The same is true in programming languages, where syntax defines how your program's *code*—keywords and other elements—are combined to create a roadmap to follow. Misplacing or misspelling keywords creates a *syntax error,* which prevents the sketch from being properly compiled. And without compiling, you can't upload the program into the Arduino.

Structure is the overall arrangement of things in a sketch. Sometimes you need to arrange code in a sketch to make it understandable to the compiler. And other times the structure is dictated by good organization and programming habits. By using a consistent structure your sketches are easier for you and other humans to understand. That helps if you or someone else needs to revise a sketch at some later date.

Anatomy of an Arduino Sketch

You know that Blink sketch you used to test the operation of your Arduino? It also serves as a good example for studying the anatomy of the typical Arduino program. If it's not already in the Arduino IDE, open the Blink sketch by choosing *File | Examples | Basics | Blink.*

- All Arduino sketches have at least two parts, named *setup* and *loop.* These are called *user-defined functions,* and they appear in a sketch like this:

```
void setup() {
}
void loop() {
}
```

- The (and) parentheses are for any optional *arguments* (data to be used by the function) for use in the function. In the case of *setup* and *loop,* there are no arguments, but the parentheses have to be there just the same.
- The { and } braces define the function itself. Code between the braces is construed as belonging to that function—the braces form what's referred to as a *code block.* There's no code shown here, so the braces are empty, but they have to be there anyway.
- The *void* in front of both function names tells the compiler that the function doesn't return a value when it's finished processing. Other functions you might use, or create yourself, may return a value when they are done. The value can be used in another part of the sketch.
- The *setup* and *loop* functions are required in every sketch. Your program must have them, or the IDE will report an error when you compile the sketch.

They're called user-defined functions to differentiate them from the functions built into the Arduino environment. The *built-in* functions are defined in a similar way, but in files you don't see when you're working in the Arduino IDE.

The *setup* and *loop* functions are the minimum your sketch needs, but you can add more. I provide an example of how to do this later in this chapter, and it's a subject revisited many times throughout this book.

Hands-on Example 1: Ye Olde LED Flasher

The Blink sketch is a good example of the trusty "Hello world!" demonstration, used to show basic functionality of a program. Instead of a video monitor to display results, for the Arduino a light-emitting diode provides enough visual feedback to validate that the sketch is working. The blinker is a good first test because the Arduino has an LED already built into it.

Being simple and to the point, the Blink example is also a good one to deconstruct, to see how it works. It's a good way to learn how to create your own sketches. Refer to Figure 2-7 to start.

The Blink program begins with a human-readable comment:

```
/*
  Blink
  Turns on an LED on for one second,
  then off for one second, repeatedly.

  This example code is in the public domain.
*/
```

This is an example of the *block comment*. The block begins with the characters /* and ends with */. Everything in between is a comment, and is ignored by the compiler. It's only there for human consumption.

Next comes the *setup* function, which you already learned about:

```
void setup() {
  // Initialize the digital pin as an output.
  // Pin 13 has an LED connected on most Arduino boards:
  pinMode(13, OUTPUT);
}
```

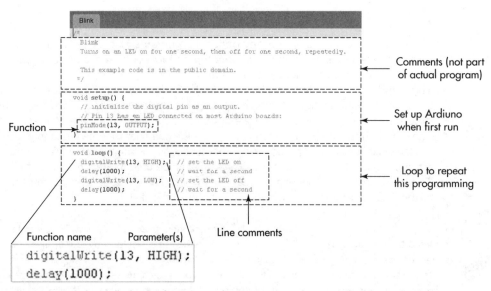

Figure 2-7 The Blink sketch, deconstructed into its principal parts. All Arduino sketches contain the two main building blocks shown in this sketch, the *setup* and the *loop* functions.

The first two lines of the *setup* function are also comments. These are *line comments,* and they extend only until the end of the current line.

pinMode is an example of a built-in Arduino function. This function requires two pieces of information, called *parameters*: a pin number—in this case digital pin 13—and whether the pin is an INPUT or an OUTPUT.

INPUT and OUTPUT are examples of keywords; specifically, they are the names of variables that have already been defined within the Arduino IDE. Recall that variables store pieces of information. INPUT stores the value 0 (zero); OUTPUT stores the value 1. The words INPUT and OUTPUT are preferred over 0 and 1 because they're more descriptive.

But why are they all-caps? This is by convention. The variables INPUT and OUTPUT are more classically referred to as *constants;* that is, the content of these variables does not change over the execution of the program. Spelling them with all-caps tells us these are constants and their value will never change.

You don't have to use all-caps for constants. It's up to you. I find lots of all-caps in a sketch make it harder to read. So I often use a mix of capitalizations. (I try to be consistent—I really do!—but no one codes their programs with one single style.)

Besides all-caps, the other common capitalization scheme used in Arduino programming is *camelCase*: high in the middle, low at the ends. The camelCase concept is simple: As variables must always be one word, you can combine words to make more descriptive names. Capitalize the first letter starting with the second word in the name. So a variable that controls a red light might be named *redLight*.

Next comes the all-important *loop* function, which repeats indefinitely—or at least as long as the Arduino is powered. The Blink sketch contains just four lines of code in the loop:

```
void loop() {
  digitalWrite(13, HIGH);    // set the LED on
  delay(1000);               // wait for a second
  digitalWrite(13, LOW);     // set the LED off
  delay(1000);               // wait for a second
}
```

The first line is *digitalWrite,* yet another function built into the Arduino. It's used to change the digital output of the pin off or on. (Note that this function is followed by a line comment, used to provide a brief explanation.) The constants *LOW* and *HIGH* represent the values 0 and 1, respectively. Like with INPUT and OUTPUT, LOW and HIGH are used to descriptively define the value you want for the pin.

The *digitalWrite* function needs two parameters (also called *arguments*): the pin number—in this case, pin D13 again—and whether the pin is to be made LOW or HIGH. The state of this pin determines if the LED is off or on, as shown in Figure 2-8.

- Making the pin HIGH turns the LED on.
- Making the pin LOW turns the LED off.

The *loop* function also uses the built-in *delay* function. This function makes the Arduino wait for a specified period of time. The value given for each delay statement is in milliseconds, or thousandths of a second. A delay of 1000 is one second.

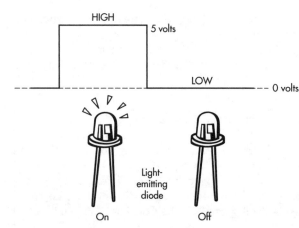

Figure 2-8 The state of digital pin 13 on the Arduino controls the operation of its built-in light-emitting diode. When the pin is HIGH, voltage is applied to the LED, and it turns on. When the pin is LOW, voltage is removed, and the LED turns off.

Notice that all the statement lines end with a semicolon. This is an important syntactical element to Arduino sketches—and, for that matter, all C-based programs. The semicolon indicates the end of the statement. The Arduino compiler needs this *statement terminator* to help it break down and analyze your code. If you miss a statement terminator, the IDE will indicate an error during compiling. You'll have to find where the mistake is, and fix it.

Hands-on Example 2: Ye Olde LED Flasher, Take 2

Blink is a great getting-started example, but it doesn't demonstrate adding external hardware to your Arduino, and that's an important aspect to microcontroller programming. It's easy enough to modify the Blink example to light an LED that's attached to another digital I/O pin.

Because of the way LEDs work they need a resistor to limit the amount of current—akin to the amount of water flowing through a pipe—that passes through them. Without the resistor, which is a common and inexpensive electronics component, the LED will burn out in seconds. Refer to the connection diagram in Figure 2-9.

 Unplug the USB cable to your Arduino before hooking things up to it. Never attach circuit wires to the Arduino while it is powered. An accidental short circuit could damage it.

Use the solderless breadboard with a red LED, a 470 Ω resistor (Ω means *ohms,* a common unit of measure for resistors), and two wires. The resistor can be plugged in either direction, but it's important to observe proper orientation of the LED. Note the flat side of the LED and how it connects in the circuit. Nothing bad will happen if you plug in the LED backwards, but it won't emit any light when you turn the circuit on.

 Most resistors are marked with a color code to indicate their value. Standard value resistors have four bands, with the first three bands indicating the value of the component, in ohms. A 470 Ω resistor uses the colors yellow-purple-brown.

The fourth band indicates the tolerance of the resistor, which is the percentage that the actual value of the component can vary above or below its indicated value. A resistor with 5 percent tolerance has a gold fourth band; a 10 percent tolerance resistor has a silver fourth band.

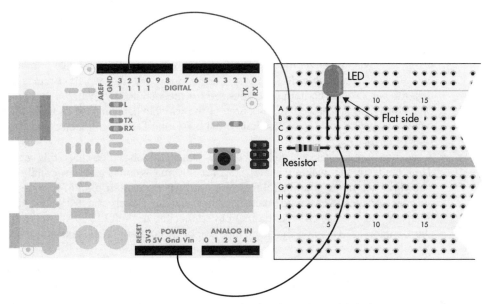

Figure 2-9   Breadboard connection diagram for the alternate LED blink demonstration. This setup shows how to properly connect a light-emitting diode to the Arduino, using a resistor to limit the amount of current flowing through the LED. Without the resistor the LED will quickly burn out.

Modify the Blink sketch so that it flashes the LED connected to pin D12 (digital pin 12) instead of the built-in LED on pin D13. Here's the altered sketch, showing a method to refer to the digital I/O pin using a variable. The sketch also demonstrates using a variable to store the delay time. (Note that this version of Blink is sans comments.)

```
int ledPin = 12;
int delayTime = 500;

void setup() {
  pinMode(ledPin, OUTPUT);
}

void loop() {
  digitalWrite(ledPin, HIGH);
  delay(delayTime);
  digitalWrite(ledPin, LOW);
  delay(delayTime);
}
```

After connecting the LED and resistor as shown, and double-checking your work, reconnect the Arduino to its USB cable. Revise the Blink sketch as shown, then compile and upload it. This time around, the external LED connected to pin D12 should flash on and off at half-second intervals.

This version of Blink also differs from the stock example in three important ways:

- The digital pin used for lighting the LED is defined in a variable, rather than directly within the code (the direct method is called a *literal*—the value is literally printed right then and there). This *variable declaration* is done in the first line, which contains the type of data the variable will contain. The keyword *int* means integer, a whole number composed of 16 bits. This is followed by the name of the variable (*ledPin*) and its value (*12*). As you may have guessed, 12 specifies that Arduino I/O pin D12 is connected to your LED and resistor circuit.
- A second variable is used to define the delay between flashes. I've called this variable *delayTime*, and stuffed it with a value of *500*, for 500 milliseconds (half a second).
- The variables, rather than literal text, are used within the *setup* and *loop* functions. This approach makes it easier to make changes to your sketch and try different values for things. You can easily change the I/O pin connecting to the LED by altering only the variable definition. To change the delay time, modify the value for the delayTime variable.

Important! Capitalization of keywords matters in an Arduino sketch—indeed, in any program that uses C syntax. You can't mix-and-match case. That means *ledPin* is different from *LEDPin*, *ledpin, LeDpIn,* and any other variation. The same goes for the delayTime variable. The sketch will not compile if you create a variable named *delayTime* and then try to use it by spelling it *DelayTime.*

For both variables I've used camelCase, which provides consistency and helps avoid capitalization mistakes. Because I've settled on CamelCase (oops, camelCase!) I know I've probably made a mistake if I later type *delaytime, DeLayTime,* or some other variation.

Hands-on Example 3: Reacting to a Pushbutton

This hands-on example demonstrates a few fundamental Arduino concepts useful in any robotics development—that is, reading a sensor and providing visual feedback. I've taken one of the examples that comes with the Arduino IDE and modified it slightly to conform to the style we'll be using throughout this book. It uses a small momentary pushbutton switch, along with that LED you added in the previous example.

I'm using a tactile or "tact" switch, which can plug into a solderless breadboard. Many other types of switches can't simply be inserted into a breadboard. See the parts list at the start of this chapter for an example tactile switch you can use. You're not limited to just this one; this type of switch is commonly available from a number of sources.

Refer to Figure 2-10 for the wiring hookup. I also provide a *schematic* view of the circuit in Figure 2-11. The schematic uses simple symbols and lines to indicate how the components are connected together. Each type of component is shown with its own (usually self-descriptive) symbol, and the lines serve to show what goes where.

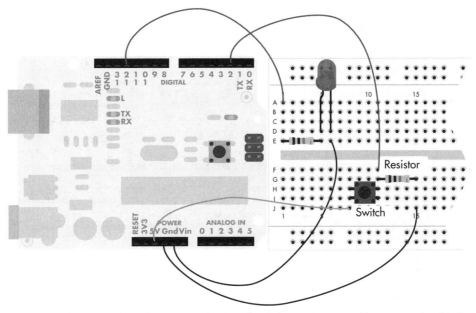

Figure 2-10 Breadboard connection diagram for demonstrating a pushbutton press and lighting up an LED. The button is a breadboard-compatible tactile momentary pushbutton switch. See the parts list at the beginning of this chapter for details.

If schematics are new to you, you'll want to bone up on how to read them. There are numerous resources for learning how to read electronic schematics. I provide a basic introduction in my book, *Robot Builder's Bonanza, Fourth Edition* (McGraw-Hill, 2011). Check at your local library for any good books on the subject; schematics for electronics haven't really changed in years, so even an old book will do.

And be sure to refer to the ARB Support Site for a list of online resources for learning about electronics and schematics. See Appendix A, "ARB Support Site," for more information.

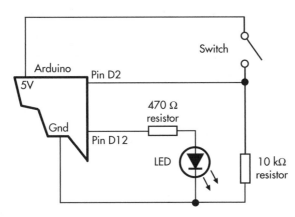

Figure 2-11 The same connection as shown in Figure 2-10, but in electronic schematic form. The symbols are the electronic components of the circuit; the lines are the connecting wires between the Arduino and the solderless breadboard.

Take a look at the wiring for the button switch. It uses a 10 kΩ resistor (10,000 ohm—brown-black-orange). This type of connection for a resistor is very common in electronics. It forms what's known as a *pulldown resistor.* It assures that when the switch is not being depressed, there is a dependable logic value applied to pin D2.

```
const int buttonPin = 2;      // Tact switch connected to pin D2
const int ledPin =  12;       // LED and resistor connected to pin D12

int buttonState = 0;          // Variable for reading pushbutton value

void setup() {
  pinMode(ledPin, OUTPUT);    // LED is an output
  pinMode(buttonPin, INPUT);  // Button switch is an input
}

void loop(){
  buttonState = digitalRead(buttonPin); // Read button state
  if (buttonState == HIGH) {            // If button is pressed
    digitalWrite(ledPin, HIGH);         //   then turn LED on
  } else {
    digitalWrite(ledPin, LOW);          //   otherwise turn LED off
  }
}
```

Here's how the program works: The first two lines set *variable constants,* so hardware connected to the various I/O pins can be referred to by name, and not pin number. This is merely for our convenience. This syntax varies slightly from the previous example: I've added a *const* keyword; *const* means constant.

 Using the *const* keyword for variables won't change how the program functions, but it saves a little bit of memory inside the Arduino when the sketch is run. The *const* keyword doesn't actually create a variable at all; rather, it tells the compiler to automatically replace all the instances of the "variable" name with its literal value. From a programming standpoint, *const* provides the ease-of-modification feature of variables, yet doesn't consume the memory that a variable does.

Another variable, *buttonState,* is defined to hold the current value of the switch, which will be a value of 0 (equivalent to LOW, or not pressed) or 1 (HIGH, or pressed).

The *setup* function specifies the *direction* of the two pins used in the sketch. The direction of the LED pin is an OUTPUT, because it controls something. The direction of the switch pin is an INPUT, because it reads the value of an input device—in this case, a switch.

The *loop* section is automatically started the moment the program has been downloaded to the Arduino. The looping continues until the board is unplugged, the Reset button on the Arduino is pushed, or a new program is loaded into memory. The loop begins by reading the status of pin D2 and storing that in a variable named *buttonState.*

- If the value is 0 (LOW), the switch is not pressed (it's "open").
- If the value is 1 (HIGH), the switch is pressed (it's "closed").

This test is done with an *if* statement, which reads

```
if (buttonState == HIGH)
```

The *if* statement performs a comparison by testing if buttonState is 1, or HIGH. (Note the use of the *double-equals* symbols... this is *very* important!) Should this statement be true, the action immediately following is carried out—the LED is turned on. Should the statement be false (buttonState is equal to 0, or LOW), the LED is turned off.

Each time you press and hold the button the LED lights up. Release the button and the LED extinguishes.

Hands-on Example 4: Making Your Arduino Sound Off

The fourth and final hands-on programming example is how to get your Arduino to make noises. Making sound is easy to do with a miniature piezo speaker element, connected as shown in Figure 2-12 (see Figure 2-13 for a schematic view). Don't expect booming sound output, but it's enough for you to hear the tones the Arduino is generating.

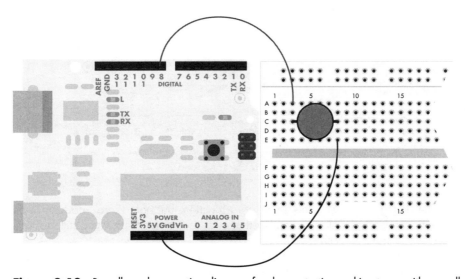

Figure 2-12 Breadboard connection diagram for demonstrating making tones with a small piezo speaker element

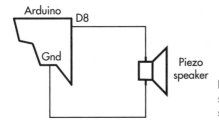

Figure 2-13 The same connection as shown in Figure 2-12, but depicted in schematic form

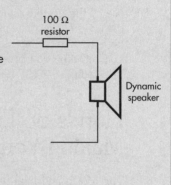

Be sure to use a piezo speaker, and not a regular dynamic (magnet) speaker. Piezo speakers have a much higher impedance—impedance is a measure of resistance to electrical current. The higher impedance means piezo speakers don't take as much current from the Arduino's I/O pins as a dynamic speaker.

If all you have in your parts bin is a dynamic speaker, you can still use it, as long as you put a 100 to 150 Ω resistor in-line with the speaker. The resistor acts to limit the current drawn by the speaker, ensuring it's a level that's safe for the Arduino to deliver. The downside to using a dynamic speaker is that the sound won't be as loud as the piezo speaker.

Sketches for generating tones can be simple to complex, depending on how fancy the music is. The Arduino IDE comes with a tone "melody" example, but the one that follows demonstrates the concept just as well, and is easier to follow and understand.

```
void setup() {
}

void loop() {
  makeSound();
}

void makeSound() {
  tone(8, 440, 500);      // Concert pitch 'A', 1/2 second
  delay(500);
  tone(8, 880, 500);      // Octave higher
  delay(1000);            // Wait 1 second
}
```

The sketch begins with the usual *setup* function, but note there's nothing to do here, so the function is empty. (Just remember that even though it's empty, you still must have a function named *setup* in each sketch.)

The *loop* function contains a reference—or *call*—to a user-defined function, named *make-Sound*. This demonstrates two things:

User-defined functions   Most such functions are pretty simple, like *makeSound*. They're just like the required *setup* and *loop* functions, but, of course, with a different name. You can use camelCase for function names.

Simplifying the structure of the loop   With more sophisticated sketches it's not unusual for your programming code to span many pages. It can be hard to follow the logic and flow of your sketch if everything is tossed into the *loop* function. You can instead separate the code into definable blocks; each block can go into a separate user-defined function. When you do this just be mindful of something called variable scope, which defines how variables created in one function can be shared with other functions.

While we're on the subject of names, both user-defined variables and functions must always be single words. They should be descriptive, and they can't start with a digit.

Variable and function names should only contain the following: letters—uppercase and/or lowercase—numbers, and the underscore. Variable names must start with an alphabetic character; they cannot begin with a number or underscore.

Good	No Good
myVariable	1Variable
good_Robot	bad-Robot
noSpacesAllowed	No Spaces Allowed
noSpecialCharacters	no&Special$Characters

The *makeSound* function combines a pair of statements, repeated over and over again, to make a really irritating high/low siren-like sound. It begins with the *tone* statement (a command provided as part of the Arduino IDE), which expects to be given either two or three parameters:

- The number of the pin connecting to the piezo speaker. In this case, it's pin D8.
- The pitch of the tone in Hertz (or Hz, cycles per second). A pitch of 440 Hz is concert A, the A note above middle C on a piano. You'll notice a second *tone* statement two lines down, this one with a pitch of 880 Hz. This is also an A note, exactly one octave higher.
- The optional third parameter of the *tone* statement is the duration, stated in milliseconds. That duration is 500, which is 500 milliseconds, or half a second.

The sketch plays both 440 Hz and 880 Hz A's for half a second, before waiting one second and repeating. This is repeated ad infinitum/ad nauseum/ad absurdum until you unplug the Arduino, rip out the speaker, go crazy, or get sick of overused Latin phrases—whichever comes first.

Chapter 3

Building the Teachbot Learning Platform

The *Teachbot* learning platform is a low-cost, 7" diameter servo-driven robot base, ready for expansion. It's called "Teachbot" because it's designed to help teach robot fundamentals using the Arduino microcontroller development board. If you're just starting out in robotics and the Arduino, this is where you want to be.

There are two versions of the Teachbot. This chapter introduces the basic version, which uses a pair of low-cost radio control (R/C) servo motors to scoot the robot across the floor. A modified version of the Teachbot, covered in Chapter 9, details exchanging the servos with high-speed gear motors, among other topics.

To keep things straight, I'll call the Teachbot introduced in this chapter—and expanded upon in the following three chapters—*Teachbot Servo*. And I'll call the modified version that uses the high-speed DC gear motors *Teachbot DC*.

Both approaches have their place, and each demonstrates a unique method of motor control. But the servo motor version is easier and less expensive to build, so that's where we'll start.

FYI

I'm starting you off head-first in the raging waters of robotics because I believe in learning by doing. As you build the Teachbot Servo you'll learn numerous important robotics concepts along the way.

Though not a requirement to enjoy this book, if you'd like even more detail, check out my book, *Robot Builder's Bonanza, Fourth Edition*. Be sure to also reference the *ARB Support Site*, listed in Appendix A.

Introduction to the Teachbot Servo Platform

The Teachbots (Servo and DC) are designed to teach robotics concepts. They're meant to allow easy exchange of components as you explore different types of sensors, electronics, and mechanical components. Instead of soldering things into a circuit board, both versions of the Teachbot use solderless breadboards to permit *plug-and-play* experimentation.

Figure 3-1 shows the completed Teachbot Servo, ready for action. This is the "all-in-one" version that shows the robot with several options, including a third deck for mounting additional electronics, and a switch for turning the robot's subsystems on and off. The version you build doesn't have to include all these features.

I built the reference design of the Teachbot Servo using 1/4" expanded PVC plastic. This plastic, which is commonly used to make indoor and outdoor signs, is inexpensive, easy to cut, and easy to drill. You can buy it mail order or through local plastics retailers and some sign-making specialty shops.

You can also use wood to construct your Teachbot (1/4" hardwood plywood from your nearby crafts store is best), acrylic plastic, foam board, picture frame mat, or most anything else that is rigid enough for all the components. Because of the extra weight, stay away from any material more than 1/4" thick, as well as ordinary softwood plywood—the kind at your neighborhood home improvement outlet—particle board, or metal.

- The *bottom* "*deck*" is a place for mounting the motors and batteries. It's a 7" diameter round disc with square cutouts for where the wheels go, and holes for mounting motors and other components. The bottom deck serves as the overall base for the entire robot. Think of it as the Teachbot's backbone.

Figure 3-1 The completed Teachbot, with Arduino microcontroller board, solderless breadboard, servos, wheels, and all body parts

- Two *modified* R/C servo motors provide propulsion. They're modified in that they allow continuous rotation; typical R/C servos turn only in a limited half-circle. You can purchase R/C servos already modified for continuous rotation, or do it yourself. Modification steps are provided for several popular models later in this chapter.

- Two *wheels* with rubber tires allow the bot to be used on a variety of surfaces, including carpet, wood, and kitchen tile. The wheels are about 2-1/2" in diameter, and measure 1/2" wide.

- *Balancing skids* fore and aft (that's fancy talk for "front" and "back"), for keeping the robot level. Some robots of similar design use wheel casters for maintaining balance, but to keep costs down the Teachbot uses two height-adjustable skids. The skids have rounded bottoms and act like small rollers.

- *Two batteries*, a 9-volt battery for powering the Arduino and other electronics, and a 4-cell AA or AAA battery holder for providing juice to the servo motors. You can use alkaline or rechargeable nickel-metal hydride (NiMH) batteries for either or both, but understand that if you use rechargeables for the servo batteries the robot will move a little slower. Batteries and their selection are covered in much more detail in Chapter 4, "Programming the Teachbot: Making It Move."

- Optional *control switch* for providing power from the two sets of batteries. The switch is wired so that you can turn everything off, apply power to just the Arduino and electronics, or apply power to both the Arduino and the motors. The switch is optional, but you'll find it handy.

- Four *deck risers* for adding a second deck—a second story if you think of the Teachbot as a building. The risers are made from ordinary 6-32 machine screws and plastic aquarium tubing.

- *Second deck*, measuring 7" round (therefore giving you about 38 square inches of space), upon which goes an Arduino Uno (or similar) development board, and two *solderless breadboards*. I've selected the kind of breadboard with 400 tie-points and removable power buss rails. These types of breadboards are common, and they're inexpensive, costing under $5 each. Between the two you'll have plenty of room for expansion.

- Optional *third deck,* also 7" diameter, in case you want to mount even more electronics to your Teachbot. The third deck is an exact duplicate of the second, and so are the four risers made of machine screws and fish tank hose.

The basic Teachbot Servo costs under $80 to build, even less if you already have some of the components, like breadboard, jumper wires, and battery holder.

So What Does It Do?

The Teachbot Servo is made to allow nearly unlimited experimentation. If it'll fit on the solderless breadboards or on the third deck, you can add it to your Teachbot.

Besides learning how to build an expandable robot base, you'll use the Teachbot Servo to discover five core robot concepts:

Motor control   Every mobile robot wants to go from one place to another. That's usually done with motors. You'll learn how to program the Arduino to control the Teachbot's twin motors to move the robot in any direction, spin, reverse, and stop. You'll also learn how to control the speed of the motors, from pedal-to-the-metal to a slow crawl.

Tactile feedback using mechanical switches   Robots need to sense the world around them in order to modify their actions—that's what makes them robots, rather than mindless machines. Small switches can be used as simple touch sensors; they work a little like the whiskers on your cat. You'll learn how to check the condition of one or more switches, and react to when the robot makes contact with something.

Changeable behaviors using light-dependent sensors   *Reactive control* is when a robot does something in response to a stimulus. A great—and inexpensive—way to experiment with this concept is making your robot sensitive to changes in light. You can completely change the behavior of your robot by simply altering a few lines of code in your Arduino sketch (remember, *sketch* is what we call an Arduino program).

Track (line, maze) following   Your robot can navigate a planned course using nothing more than a pair of simple sensors that point toward the ground. Sensing a track is a classic application in robotics, as it teaches how to read sensors, correlate their respective values, and modify the action of the robot's motors to compensate. Track following demonstrates the value of the *algorithm,* a core concept in programming and in robotics control.

Explore surroundings   Armed with one or more distance or proximity sensors, your robot can intelligently navigate a room, looking for things to explore. Such sensors let the robot see inches and even feet ahead, providing rudimentary "eyesight." By altering the programming in your robot you can have it advance toward any object it sees, yet avoid direct contact. Or if you're playing combat war games with another bot, intentionally make hard and fast contact.

The Teachbot Servo is not limited to just these five tasks. Once you've completed the Teachbot Servo projects described in this book you are encouraged to continue with your own experiments. The whole idea of the Teachbot is that it's a springboard for learning, a never-ending classroom in all things robotics.

Making the Teachbot Servo Base

The Teachbot Servo is constructed with four body pieces, held together with hardware fasteners. Table 3-1 provides a full list of mechanical parts, and Table 3-2 gives you a parts list for the motors, wheels, and electrical components.

All body pieces assume 6mm or 1/4" thick material. As I mentioned, my prototype used 6mm expanded PVC plastic, but you can select your favorite material instead.

Cut the parts using the tools of your choice. Because the Teachbot decks are round, you'll have the best luck using a motorized jig saw (for plywood), motorized scroll saw (for plywood or PVC), or hand scroll saw (for PVC). For intricate cuts you'll want to use a hand scroll saw, as that will give you more control.

FYI   As a service to readers, the base parts for the Teachbot are available precut and predrilled in case mechanical construction isn't your bag. See Appendix B, "Parts Connection," for more details.

If using foam board or picture mat, you can cut the pieces using a sharp hobby knife or mat cutter. The usual safety precautions apply. A circle-cutting jig makes perfect circles when using these materials. If you don't own a circle jig, see if the local picture frame store will make the cuts for you. When using picture mat material cut two of everything, and double-up the pieces for extra stiffness.

Table 3-1	Teachbot Servo Mechanical Parts
Quantity	**Description**
1	7" diameter bottom deck, with wheel well cutouts for the drive wheels.
1	7" diameter top deck.
2	Servo mounts.
4	90° plastic L brackets for attaching the servo mounts to the bottom deck. These brackets measure 3/4" × 3/4", with hole centers at 3/8", and are made to work with the two servo mounts.
16	4-40 × 1/2" machine screws and nuts for attaching the servos and servo mounts to the bottom deck.
4	Deck risers consisting of: (4) 3" 6-32 machine screws, (8) 6-32 nuts, and (4) poly tubing cut to length to act as spacers (see text).
2	Skids consisting of: (2) 6-32 × 3/4" machine screws, (4) 6-32 hex nuts, and (2) 6-32 acorn (cap) nuts.
3	Sets of mounting hardware for Arduino Uno, consisting of (6) 4-40 × 1/2" machine screws and (3) 1/2" or 3/4" length plastic 4-40 threaded standoffs.
Misc	Hook-and-loop (Velcro) strips for mounting battery holders and solderless breadboard and small pieces of double-sided foam tape.

Table 3-2	Teachbot Servo Motors, Wheels, and Electrical Parts
Quantity	**Description**
2	Standard size R/C servo motors, modified for continuous rotation.
2	2-1/2" or 2-5/8" diameter wheels, with hubs to attach to the servo motors.
1	Arduino Uno (or compatible) microcontroller board, with USB programming cable.
2	Solderless breadboards, 400 tie-points, separable power rails (see text).
1	Set of solderless breadboard wire jumpers (or make your own using 22-gauge solid conductor wire).
1	Length of 12 (or more) breakaway 0.100" male header pins, double-sided (long) pins; see text.
4	AA alkaline or nickel-metal hydride rechargeable batteries.
1	AA × 4 battery holder, with female header connector; see text.
1	9-volt battery.
1	9-volt battery clip, with 2.1mm polarized barrel plug; see text.
1	9-volt metal or plastic battery holder.

CUTTING OUT THE PIECES

Begin by cutting out the body pieces for your Teachbot.

- The *bottom deck* measures 7" diameter, and has so-called *cutouts* for the wheels (see Figure 3-2). The deck includes a number of holes, of which only eight are required. Four of the holes are for mounting the servos; the four other main holes are for the risers that separate the two decks. Any other holes are up to you. I've included several additional holes at the front and back of the deck for mounting bumper switches and other sensors. The wheel cutouts measure 2-5/8" by 5/8", sized for commonly available 2-1/2" or 2-5/8" diameter robotic wheels for R/C servo motors.
- The *top deck* also measures 7" in diameter (see Figure 3-3). Only four of its holes are critical; these mate with matching holes in the bottom deck using a set of four *risers* (also called *standoffs*). One or more 1/2" diameter holes provides thruways for wires from the bottom deck. The exact size and placement of these holes are up to you.
- The two *servo mounts* (see Figure 3-4) attach the servos to the bottom deck. You can make these yourself, or if you choose, purchase them separately from the sources provided in Appendix B. If you make the mounts, be aware that *sizing is critical.* The two holes on either side of the mount must be spaced 3" apart to accommodate the same hole spacing in the bottom deck.

Except for the large holes in the center of the top deck, all holes are drilled with a 9/64" bit.

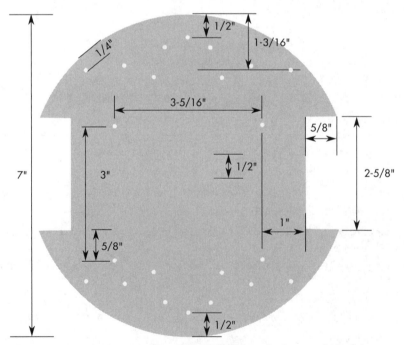

Figure 3-2 Layout pattern for cutting and drilling the bottom deck of the Teachbot. The only truly critical dimensions are the cutouts for the wheels and the placement of the two sets of holes immediately beside the wheel cutouts. These holes are for the servo mounts. See Figure 3-6 for a description of all holes.

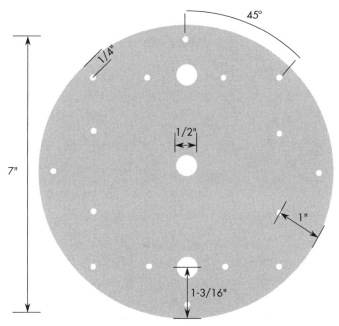

Figure 3-3   Layout pattern for cutting and drilling the top deck of the Teachbot. Critical holes are the four small ones nearest the outside. These must match the four servo mounting holes in the bottom deck.

HOLE ALIGNMENT

Figure 3-5 indicates the holes that need to be in alignment between the bottom and second decks. These are the holes that serve as the risers between the decks. After cutting out the two decks, you may want to mark the four holes in the second, sandwich the two decks together,

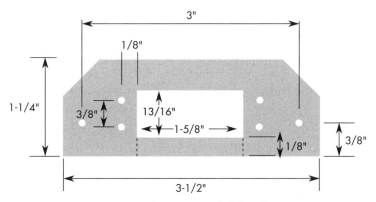

Figure 3-4   Layout pattern for cutting and drilling the servo mounts. You'll need two of these. If cutting the inside rectangle proves difficult, you can instead make the mounts by cutting through at the dotted line. The mount will be a little more fragile, so handle it carefully. Use all four screws to secure the servo in the mount, rather than just two.

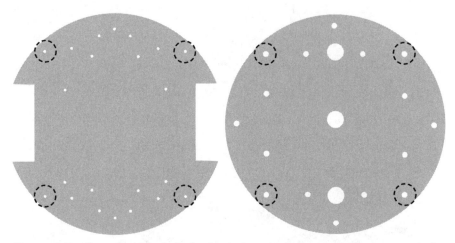

Figure 3-5 Placement of the holes for the deck riser is fairly critical. These are marked as shown for the bottom and second decks.

and drill the holes as a complete unit. That way, if your marking or drilling is slightly off, the holes will still line up.

ADDITIONAL OPTIONAL DECKS

Teachbot is designed for expandability, and should the twin decks not provide enough space for all your experiments, you can add more decks. Each additional deck is merely a duplicate of the second.

I don't recommend any more than three decks total. Why? The more decks you add, the more weight that gets piled onto the bot. After a certain point the weight is too much for the motors to move. An overly tall robot is also prone to instability problems.

Assembling the Teachbot

With the body pieces cut and drilled, and all other parts in hand, you're ready to build your Teachbot. Here's how.

STEP 0—PREPARE THE BASE PARTS

Before assembly you may wish to use 150-grit sandpaper to smooth the edges of the base parts.

Orient the bottom deck so that the holes are aligned as shown in Figure 3-6. Note that the holes for each servo are not symmetrically placed on the deck. This is to accommodate the offset of the servo drive shaft. While there is technically no "front" or "rear" of the Teachbot, for the purposes of assembly and initial use, the top of the illustration in Figure 3-6 is the front, and the bottom is the rear.

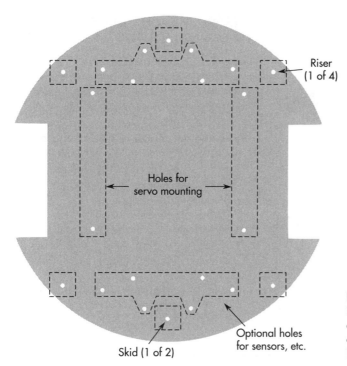

Figure 3-6 Explanation of the holes in the bottom deck. The holes marked as optional are for sensors, switches, and other hardware, and their exact location is not critical.

STEP 1—ATTACH SERVOS TO SERVO MOUNTS

Insert a servo into a servo mount by sliding it back-end first through the mount. The fit may be tight, depending on the make and model of the servo. (As necessary, enlarge the rectangle for the servo using a file or coarse sandpaper.) Do not force the servo into the mount, or the mount may be damaged.

Secure the servo to the mount with 4-40 × 1/2" screws and hex nuts (Figure 3-7). You can use four screws for each servo, or only two. When using two screws, position them on opposite corners of the servo mounting flange, as shown.

Repeat for the opposite servo and mount. *Be sure to construct the second servo and mount in mirror-image to the first!* Refer to Figure 3-11 to see how the motors should be inserted into the mounts.

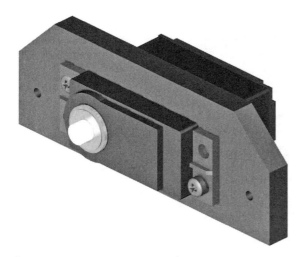

Figure 3-7 Servo motor secured into a servo mount. You need two of these.

Figure 3-8   Attach two L brackets to the servo mount. The L brackets should be flush with the bottom of the servo mount.

STEP 2—ATTACH BRACKETS TO SERVO MOUNTS

Using 4-40 × 1/2" machine screws and nuts, attach two plastic L brackets to each of the servo mounts (Figure 3-8). You'll make a "left" and a "right" mount assembly.

- For the left mount assembly, the motor shaft should face to the left and toward the "top" of the deck (as referenced in Figure 3-6). Attach the L brackets to the right side of the mount.
- For the right mount assembly, the motor shaft should face to the right, also toward the top of the deck. Attach the L brackets to the left side of the mount.

Insert the machine screws through the L bracket, then through the servo mount. Secure on the other end with a nut. Before tightening, be sure the bottom of the L bracket is flush with the bottom edge of the servo mount.

STEP 3—ATTACH SERVO MOUNTS TO BOTTOM DECK

Attach the left mount assembly to the bottom deck using two 4-40 × 1/2" screws and nuts. The screws should come up from the underside of the deck, through the L bracket, and then into the nut, as shown in Figure 3-9. When orienting the mount assembly, be sure that the servo shaft is centered in the wheel well cutout. Otherwise, the wheel won't fit onto the motor without hitting against the deck.

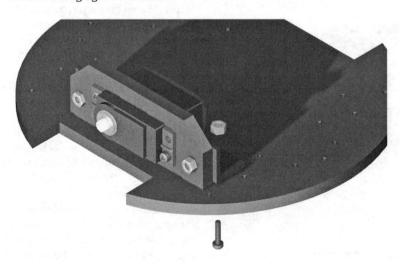

Figure 3-9   Secure the servo mounts to the bottom deck using machine screws and nuts.

Align the assembly so it is parallel with the wheel well cutout, and tighten all screws. Repeat the same procedure for the right mount assembly.

STEP 4—ATTACH BALANCING SKIDS

Attach the front and rear skids as shown in Figure 3-10. Each skid uses a 6-32" machine screw, one or two hex nuts, and an acorn (cap) nut.
For each skid:

1. Thread a nut all the way to the head of the screw.
2. Insert the screw into the base from the top (the side with the motors).

3. Thread a second nut onto the screw.
4. Attach an acorn nut to the screw, and tighten the nut against the hex nut.

Repeat these steps for the other skid.
You may adjust the height of the skid by adjusting the nuts. This may be necessary if the wheels you are using are not exactly 2-1/2" in diameter. Remove the nut closest to the acorn if you need one skid shorter than the other. If you need greater height adjustment, merely use a longer machine screw.

You want to avoid adjusting the height of the skids so that they touch the ground but the wheels don't. The height of the skids should ideally allow for a back-and-forth "rocking action" of about 1/8" to 3/16" once the wheels are attached.

STEP 5—ATTACH WHEELS TO SERVOS

Attach the wheels to the servos. Each wheel is secured with a small self-tapping screw that is supplied with the servo. Note that the servo shaft is splined, and this spline matches the wheel hub. Be sure to press the wheel onto the shaft firmly while tightening the screw. Do not over-tighten the wheel mounting screw, but be sure the wheel is on snugly. Figure 3-11 shows how the wheels look on mounted servos.

Figure 3-10   Teachbot uses static skids for front-and-back balance. You can adjust the height of each skid to compensate for the diameter of wheels you use.

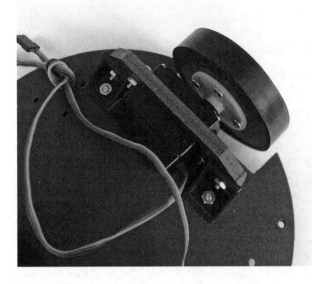

Figure 3-11 How the completed servo mount should look, with wheels attached to the servo

STEP 6—ATTACH BATTERIES

Secure the 4-cell AA battery holder to the approximate center of the bottom deck using two 3/4" squares of hook-and-loop (Velcro) to keep it in place.

Insert a 9-volt battery into a metal 9-volt battery holder. Use 3/4" square hook-and-loop piece to secure the clip next to the AA battery holder.

 Use heavy-duty hook-and-loop to keep the batteries in place.

For a more secure fit, mark a hole about 3/16" from the edge of the 9-volt battery, and drill for a 4-40 screw. Use a 3/4" long threaded nylon standoff to create a backstop for the 9-volt battery.

Note the electrical connections for both the 9-volt battery and the AA battery holder:

The 9 volt battery uses the traditional two-prong battery clip, terminated on the other end with a 2.1mm barrel plug. This plug inserts into the power jack of the Arduino. You can make this power lead yourself by soldering a barrel plug onto a standard two-prong battery clip, or purchase one ready-made. Either way, *be absolutely sure* the + (positive) connection is the center of the plug; the − (negative) connection is the outside barrel.

The AA battery holder uses a female 0.100"-spacing pin header connector (this connector is detailed more fully in the next chapter). You can use a connector with two or more pins; the additional pins can be used to help assure proper polarity.

With just two pins you must be VERY careful to never (and I mean *NEVER, EVER!*) reverse the polarity of the connector. If you do, your servos may be instantaneously and permanently damaged. By using (for example) a four-pin connector, you can block up one of the unused terminals. This helps prevent you from reversing the connector when you plug it in. (Of course, still be careful, no matter what system you use!)

Insert fresh AA batteries into the four-cell holder, and attach the clip to the 9-volt battery.

STEP 7—ATTACH ARDUINO TO THE SECOND DECK

Lay out the Arduino and the two solderless breadboards as shown in Figure 3-12. Exact placement is not critical, but keep the corners of the breadboards from overhanging the deck by more than about 1/4". Be sure to leave a space of about 1/8" around the top and bottom of the Arduino to accommodate the width of the nylon standoffs you'll use to mount the board to the deck.

Once the pieces have been arranged, carefully mark the holes for securing the Arduino. You can use just two, on opposite corners, but for a more reliable hold, plan on mounting with three screws.

After marking, drill the holes using a 9/64" bit. Secure the Arduino board to the top deck using 4-40 machine screws and standoffs. Try to use only plastic fasteners for any part that may contact the Arduino. If you don't have plastic screws it's okay to use metal, as long as you insulate the head of the screws with a plastic washer. This minimizes the possibility of a short circuit.

STEP 8—ATTACH BREADBOARDS TO THE SECOND DECK

Separate the power buss rails from at least one of the solderless breadboards as shown in Figure 3-12. Use hook-and-loop to mount the breadboards and power busses around the Arduino as shown in Figure 3-13.

Figure 3-12 The Arduino board mounted on the second deck using nylon standoffs, and surrounded by two solderless breadboards.

Figure 3-13   The Teachbot uses an Arduino Uno (or compatible; my photography model just happens to use an older Arduino Diecimila) and two solderless breadboards. The breadboards are laid out as shown. You'll be making fine-tuning adjustments to the boards in the chapters that follow.

Many breadboards come with double-sided self-adhesive tape. I recommend that you don't use the tape. Instead, mount the board using squares of hook-and-loop (Velcro). This allows you to easily remove the board when you need to.

Some solderless breadboards come with self-adhesive foam tape on the bottom. I don't recommend using this, as the tape is more permanent than Velcro or other hook-and-loop. It's easier to reposition and replace the breadboards when you use hook-and-loop fastener.

Instead of securing the boards with tape, either carefully remove the double-sided tape from the underside of the breadboards, or use the tape to first attach the breadboard to a thick piece of plastic. Then secure the bottom of this plastic to the base using hook-and-loop.

STEP 9—ATTACH SECOND DECK TO BOTTOM DECK

To complete the Teachbot, secure the top deck using homemade standoff risers. Construct the risers as shown in Figure 3-14. Make the plastic riser piece from 0.170" I.D. (inner diameter) flexible tubing. For 3" long machine screws, the tubing should be cut to a length of 2-1/4".

Start by threading a screw through the bottom of the base. Add the tubing; then thread on a nut. Tighten the nut so there is moderate pressure against the tubing. This is what gives the riser its strength. Don't overtighten or the tubing will simply buckle.

To attach the second deck, line up four of its outermost holes with the risers. Secure the deck in place using four more nuts. Tighten these nuts just enough so that they don't become loose on their own.

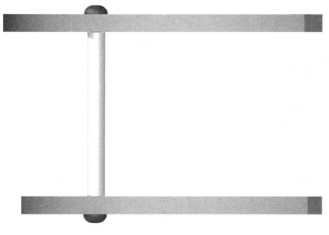

Figure 3-14 The standoffs, or risers, between the two decks of the Teachbot are made with ordinary 6-32 machine screws and rubber aquarium tubing. The tubing is used as a supporting column between the decks.

While putting the deck in place thread the leads for the batteries and servos through the center hole of the top deck. In this basic version of the Teachbot Servo power switches for the batteries are left out, though this is an option you can add at a later time. See Chapter 6, "Programming the Teachbot: Letting It Explore," for plans on how to wire a switch to control both battery sources at the same time.

Without the switch you'll need to manually plug in the batteries each time you use the Teachbot. When you're done playing, be sure to unplug the batteries to keep them from draining.

How the Teachbot Servo Moves Around

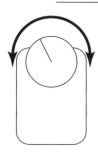

The Teachbot Servo platform uses *differential steering,* where the base is propelled by two motors and wheels on opposite sides. You can see how differential steering works in Figure 3-15. With this arrangement, the Teachbot is able to move forward and back, turn left and right, and spin in place. These movements are made possible by individually controlling the direction of each of the two motors.

To keep down costs and minimize construction complexity, the Teachbot Servo uses a pair of skids, placed in the front and rear, to provide balance. The skids are smooth and polished metal, so they present little drag on whatever surface the robot is rolling over. Even so, the Teachbot is best suited for travel on hard surfaces or carpet with a very short nap.

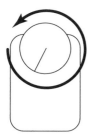

The two drive motors run off their own battery supply, a set of four AA rechargeable or nonrechargeable cells. Teachbot's motors are standard size R/C servos, and measure about 1-1/2" × 3/4" × 1-3/8". As with all R/C servos of this type the motor casing provides easy screw holes for mounting.

Where the Teachbot motors differ from regular R/C servos is this: Most servos are made to turn no more than about a half-circle one direction or another. They're used for things like steering a model car or operating the flaps in a model airplane. The motors on the Teachbot are modified so they rotate continuously.

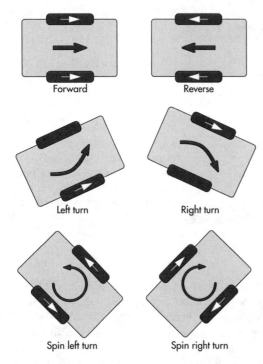

Figure 3-15 The Teachbot uses differential steering. By changing the direction of the two wheels relative to one another the robot can be made to move forward and back, or turn right or left.

I prefer using R/C servos that come from the factory already modified for continuous rotation. Why? These are easier to use because they don't require you to take them apart and hack them. For the Teachbot prototype I used a pair of GWS S-35 servos, which come ready-to-go as robot drivers. But there are other continuous rotation R/C servos available, including the Parallax Continuous Rotation Servo and the SpringRC SM-S4303R. See Appendix B, "Parts Connection," for more information on these sources.

If you prefer, you can modify regular R/C servos so they rotate continuously, as explained later. But before getting into that, let's take a short (and I mean *short*) side trip down servo lane and explain how these things are meant to work. You can't simply apply voltage from a battery and expect an R/C servo to rotate.

SERVOS 101

Take apart an R/C servo and you'll find a motor, some gears, and control electronics.

Main motor   The heart of the servo is a *DC motor* (DC = direct current, like that from a battery). It can spin in either direction.

Reduction gears   The DC motor turns at high speed, so a series of *gears* are used to reduce the output speed. In reducing the speed, the *torque* (amount of work the motor can do) is increased. (This technique is often referred to as *mechanical advantage.*)

Control circuitry   The output gear is connected to a *potentiometer,* a common electronic device similar to the volume control on a radio. As the output gear of the servo turns, so does the potentiometer; this is how a servo motor knows where it's pointing. Most potentiometers have a limited rotation. There's a raised nub in the output gear that prevents the servo from turning beyond the limits of the potentiometer.

Controlling the Position of a Servo

Connection to a servo is via three wires: power, ground, and signal.

- *Power* is the positive connection from a set of batteries (or other source). Most servos are engineered to be operated at between 4.8 volts and 6 volts.
- *Ground* is the negative connection from the batteries or other source. It's necessary to complete the power circuit.
- *Signal* is a special control voltage that determines the angular position of the servo. R/C servos are engineered to be controlled by sending them electrical pulses via this signal wire. These pulses repeat 50 times a second (50 *Hertz,* or *Hz*). The pulses themselves vary in how long they last, between 1.0 and 2.0 milliseconds in length—one millisecond is 1,000th of a second.

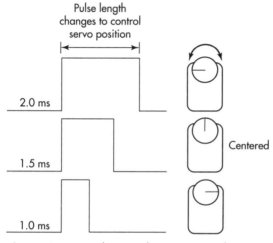

Figure 3-16   Radio control servos are made to move to a specific angular position, depending on the duration of a series of pulses.

The duration of the pulse indicates where the servo points to. At 1.5 milliseconds (ms), the servo points straight ahead, as shown in Figure 3-16. At 1.0 ms, the servo turns to its intended stopping point in one direction; at 2.0 ms, the servo turns to its intended stopping point in the opposite direction.

The special pulsed signal required by a servo is why you can't simply connect it to a battery and expect the motor to move. Without the proper pulses the motor will not turn.

Fortunately, it's not difficult to program an Arduino to provide these pulses. In fact, the Arduino comes with a ready-made library (collection of programming statements) that you can use to control one or more R/C servo motors. You'll read more about this in Chapter 4, "Programming the Teachbot: Making It Move."

The actual duration of the pulses provided to a servo can extend beyond the 1.0–2.0 ms range. The extended pulses make the servo move a little farther in either direction. The minimum and maximum pulse range depends on the servo, and trying to move a servo beyond its physical stops may damage it. But it's not unusual for a servo to allow for a range of between 0.75 ms and 2.25 ms.

In programming for robotics, it's more common to measure the pulse width in microseconds—abbreviated μs—rather than milliseconds. A microsecond is one millionth of a second, or 1,000 of a millisecond. The 1.0–2.0 ms range is stated as 1,000 to 2,000 μs.

From here on out in this book, when discussing servo pulse width I'll provide the measurements in microseconds.

Servos for Continuous Rotation

R/C servos are typically used when precise angular positioning is required. But as you've read, servos can also rotate continuously, either by design, or via modification that you perform yourself. Why use R/C servos for driving your robot? First is that they tend to be less expensive than comparable DC gearmotors of the same specification. Servos come with their own interface electronics, so you don't need any additional circuitry other than your Arduino. And finally, their design provides easy methods to mounting them.

Servos that rotate continuously act like an ordinary geared DC motor, except it's still controlled by sending it pulses:

- To make the motor go in one direction, send it 1,000 μs pulses.
- To make the motor go in the other direction, send it 2,000 μs pulses.
- To make the motor slow to a stop, send it 1500 μs pulses.
- To make the motor stop altogether, stop sending it pulses.

Note that when using continuous rotation servos there is no benefit of using pulses above or below the standard 1,000–2,000 μs range. That tidbit is only useful when using a servo for angular positioning.

FUNDAMENTALS OF SERVO MODIFICATION

The easiest way to modify a servo so it rotates continuously is to disengage its potentiometer so it no longer turns when the motor rotates. As part of this servo surgery, any physical stop on the output gear of the servo is removed. This enables the gear to spin continuously without it getting caught up on anything.

You can convert many kinds of servos to rotate continuously. But keep in mind that once modified, the servo is no longer capable of precise angular rotation.

BASIC MODIFICATION INSTRUCTIONS

Servo modification varies somewhat between makes and models, but the basic steps are the same.

1. Remove the case of the servo to expose the gear train, motor, and potentiometer. This is accomplished by removing the four screws on the top or bottom of the servo and separating the two halves of the case.
2. File or cut off the nub on the underside of the output gear that prevents full rotation. This typically means removing one or more gears so you can get to the output gear. So be careful not to misplace any parts. If necessary, make a drawing of the gear layout so you can replace things in their proper location.
3. Remove the retainer clip at the bottom of the output gear. Doing this disengages the potentiometer from the gear, so that the pot no longer turns when the gear does. On some servos there is no retainer clip; the servo engages into a spline molded into the bottom of the gear. On these you need to carefully drill out the bottom of the gear.
4. Reassemble the case.

In the following two sections you'll find detailed modification instructions for the Hitec HS-422 and the Futaba S3003. With the same or minor variations, the steps that follow can be applied to similarly designed servos.

The construction of some R/C servos makes them impractical to modify. One case in point: Modification isn't recommended if the output gear of the servo is supported only by the potentiometer shaft—there's no lower ball bearing or bushing. Such servos tend to be the least expensive and not good choices as drive motors for your robot.

Mini and micro servos are among the hardest to modify, and attempting to do so may permanently ruin the motor. You're often better off getting a miniature DC gear motor (and suitable interface electronics) and using it instead.

TOOLS YOU NEED

You'll need the following tools to complete the conversion process:

- #0 Phillips screwdriver
- 1/8" or smaller flat-bladed screwdriver
- "Nippy" cutters, X-ACTO blade, or razor saw
- Small flat jeweler's file

CHOOSING A SERVO TO MODIFY

The best servos, from a standpoint of easy modification, are those that have one or more of the following personality traits:

Use a lower ball bearing or bushing that supports the output gear   Many of the better servos use a ball bearing or metal bushing on the output shaft to provide greater mechanical support. These make the best motors to modify. Or at the very least, the output gear should be supported by a molded-in ledge, rather than directly on the potentiometer shaft. In the latter, the motor is not suitable for modification.

Use a removable potentiometer shaft clip   The clip is located underneath the output gear, and engages the potentiometer shaft. Servos that have a snap-in clip are easiest to modify. Servos that lack a removable clip will use instead a molded-in channel that the potentiometer fits into. This is the case with such budget-priced servos as the Hitec HS-311 and Futaba S-3003. If your servo is so constructed, you'll need to carefully drill out the bottom of the output gear in order to remove the spline. Instructions are provided later.

Use a plastic output gear, rather than a metal gear   The better, heavy-duty servos use a metal output gear. While the metal is able to handle heavier loads, it is significantly harder to modify, especially if you don't have the right tools. For now stick with servos with a plastic output gear.

Are normal sized or larger   Mini and micro servos are much more difficult to modify. You may ruin one or two before you get the hang of it.

STEP-BY-STEP: MODIFYING A HITEC HS-422

The Hitec HS-422 is a good "middle of the road" servo: low price, available anywhere, and well made. It employs both a top and bottom oil-impregnated brass bushing and a removable clip on the underside of the output gear.

Throughout the following, use care to avoid wiping off or absorbing through your skin too much of the lubricant used for the internal gears of the servo. If you think too much of the lubrication has been lost, you can always add more just prior to reassembly. Clear (or white) gear grease is available at any hobby store that sells R/C parts.

Important! Before performing the modification, first test the servo for proper operation. Opening the servo and modifying it voids the warranty.

These steps are virtually the same for several other popular low-cost servos, including the venerable Futaba S-148 and the S03 series of servos from GWS.

1. Using the Phillips screwdriver, remove the horn wheel, if one is attached to the servo.

2. Untighten the four casing screws from the bottom of the servo. If they don't get in the way in the following steps, you can keep the screws in place in the bottom of the case. Unscrew just enough to remove the top of the servo case. Note that on a few servos, notably the S03 series from GWS, the case screws are removed from the top, and these should be removed all the way and set aside while you work.

3. Observe the orientation of all the gears. Remove the center gear, being careful not to unseat its metal shaft. On most servos the center gear cannot be easily removed without also lifting up the output gear. Place the center gear aside.

4. Remove the output gear.

5. Using the nippy cutters, X-ACTO blade, or razor saw, remove the nub on the top side of the output gear. I like using the nippy cutters myself, *but exercise caution!* The harder the plastic, the more likely the nub will break off at high speed. Wear eye protection. Always nip first on the long side, to prevent possible breakage of the gear, and cut off only a small portion at a time. When using an X-ACTO blade or razor saw, the obvious precautions against cutting your fingers off should be observed. Work slowly.

6. Odds are no matter what cutting technique you use, a small portion of the nub will remain. This can be filed down with a small flat file.
7. Use the small-bladed screwdriver to remove the metal retaining ring from the underside of the output gear. This ring retains the potentiometer shaft clip, and also serves as a bearing surface.

8. Use the small-bladed screwdriver to remove the retaining clip.
9. Replace the metal retaining ring back into the output gear.
10. Align the potentiometer shaft so that it's centered. If needed, rotate it back and forth to find the center.

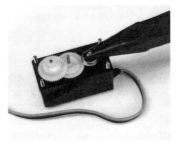

11. As an optional step, you may want to connect the servo to your control circuit. Apply 1500 µs (1.5 ms) pulses from a servo tester or microcontroller. If the motor turns (even slowly), rotate the potentiometer shaft until all rotation is "nulled out."

12. Once set to its center, you can leave the shaft as-is, or apply a *very small* dab of cyanoacrylate glue (Super Glue) to keep the shaft in place. Do not apply too much glue, or the potentiometer may be damaged.

13. Reassemble by placing the output gear on its seat over the potentiometer. Replace the middle gear, and observe that all gears properly mesh. Add more grease at this point, if needed. Finally, replace the top case and the four case screws.

Test the servo for proper operation by connecting it to your control circuit. A series of 1.5 ms pulses should stop the servo. A nominal signal of 1.0 ms pulses should rotate the servo in one direction; 2.0 ms pulses should rotate the servo in the other direction.

STEP-BY-STEP: MODIFYING A FUTABA S3003 SERVO

The Futaba S3003 is a low-cost alternative to servos with metal bushing or ball bearings. As with many servos of its type, it doesn't use a retained clip for the potentiometer. So you need to drill out the bottom of the gear so that the gear will no longer engage with the potentiometer.

1. Follow steps 1 through 6 in the previous procedure.

2. Using a spare servo horn (the larger the better), attach the horn to the output gear using the mounting screw provided with the servo.

3. For steps 7 though 9, instead carefully drill out the bottom of the output gear with a 3/16" (or thereabouts) drill bit. Remove as much of the plastic as necessary so that the potentiometer shaft will not engage the bottom of the gear. Use a drill press if you have one. See Figure 3-17. If you don't have a drill press, get someone to hold the output gear (see the following note) while you drill out the spline.

4. Complete the remainder of the steps in the previous procedure.

When you drill out the spline in the bottom of the gear, hold the gear steady by clasping the servo horn—*not* the gear—using a pair of heavy-duty pliers. Do not clamp by the shaft or gear face, as this could wreck the plastic and you'll end up with a badly functioning servo. Be sure not to ream or drill out too much, or you'll ruin the gear.

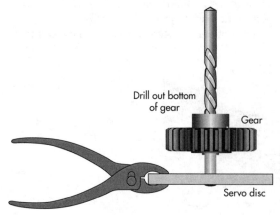

Drill out bottom of gear

Gear

Servo disc

Figure 3-17 You'll need to *carefully* drill out the bottom of the output gear on servos that don't have a retainer clip for the potentiometer. Be sure to use a pair of pliers to keep hold of the gear, as noted in the text.

As you work inside the servo you're bound to remove some of the lubricating grease inside. As needed, apply new grease. You can get clear or white grease for fine mechanical parts at many hobby stores and electronics outlets.

While it may be tempting, don't apply petroleum-based oil to the gears, such as 3-In-One oil or a spray lubricant like WD-40.

TESTING THE MODIFIED SERVO

After reassembly but before connecting the servo to a control circuit, you'll want to test your handiwork to make sure the output shaft of the servo rotates smoothly. Do this by attaching a control disc or control horn to the output shaft of the servo. Slowly and carefully rotate the disc or horn and note any snags. Don't spin too quickly, as this will put undo stress on the gears.

If you notice any binding while you're turning the disc or horn, it could mean you didn't remove enough of the mechanical stop on the output gear. Disassemble the servo just enough to gain access to the output gear and clip or file off some more.

See the next chapter for details on using continuous rotation servos with the Teachbot.

Chapter 4

Programming the Teachbot: Making It Move

Mobile robots are meant to move, and they use motors to do it. The radio control (R/C) servo motor is among the easiest to adapt to a small robot chassis—it's compact, easy to mount, and affordable. The Teachbot Servo uses two such motors, situated on each side of the robot. Simply by controlling the speed and direction of these motors you can scoot the Teachbot across the room and maneuver it with amazing accuracy.

In this chapter you'll learn how to operate the R/C servo motors on the Teachbot, for both autonomous operation and manual control. What follows assumes you've built the Teachbot according to the plans in Chapter 3. If you haven't yet done so, you'll want to complete that part before continuing.

 Be sure to visit the ARB Support Site (see Appendix A) for additional code examples and variations. All sketches from this book are available for ready download.

FYI Brand new to Arduino programming? Then be sure to first read through Chapter 2, "Arduino Up and Running." And, of course, refer to the official Arduino Language reference pages at *www.arduino.cc*.

Wiring Up and Testing a Single Servo

Let's start simply and discover how to run just one servo motor. Once that's mastered, it's a simple process to control two motors.

Recall from Chapter 3 that R/C servos already contain all the electronics needed to control the speed and direction of the motor. The good news is that you don't need any extra circuitry to connect to a servo motor and operate it. The bad news is that you cannot simply apply voltage to the motor wires and expect it to turn.

As you read in the preceding chapter, R/C servos require a special timing signal, in the form of a repeating pulse, to set their speed and direction. Ah, but good news again: The Arduino has no problem providing these signals. All it requires is that the servo be electrically connected in the proper manner, and some simple programming statements to regulate the operation of the motor.

 If you don't have a spare standard size R/C servo motor lying around the house, you can use one of the two motors you attached to your Teachbot Servo. Place the bot on a book or other platform to lift the wheels off the ground. Orient the robot so that you can observe the motor and wheels turn as you experiment.

Refer to the wiring schematic in Figure 4-1. It shows how to connect a servo motor to the Arduino. Pay particular attention to how the three wires from the servo are attached to the Arduino I/O pins. *It is absolutely critical that you don't reverse any of these wires*, or damage could result to your servo, and possibly the Arduino as well.

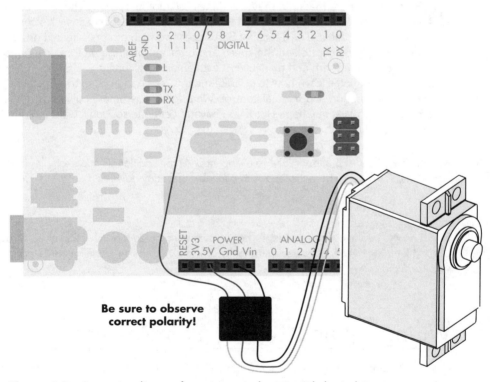

Figure 4-1   Connection diagram for testing a single servo with the Arduino

Servo motors use color-coded wiring to indicate the meaning of each wire, but there's no universal standard for the colors used. Many servos use one of several color schemes, and the following two are the most common. Here, *Ground* is 0V, and *Power* is 4.8 to 6 volts (some servos can handle higher voltages, but that's not a concern for right now).

Wire Colors	Meaning	Connect to Arduino Pin
Black	Ground	Gnd
Red	Power	5V
White (or Yellow)	Signal	Digital 9
Brown	Ground	Gnd
Red	Power	5V
Orange (or Yellow)	Signal	Digital 9

Except for some old servos and specialty servos, the Power wire is always in the center. The color of this wire is usually red. The Ground wire is most often black or brown. When in doubt, check the specifications for the servo.

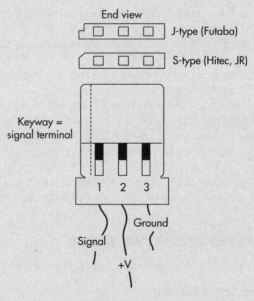

Many servos use a polarized connector to prevent you from plugging it backwards into an R/C receiver. Look closely and you may see a notch or groove on one side of the connector. On virtually all servos this notch or groove marks the Signal wire.

PROGRAMMING THE SERVO

You're now ready to program the servo to make it move. Open the Arduino IDE program, and type the following program. (You may also download it from the ARB Online Support site; see Appendix A for details.) Be sure to type it exactly as shown, including spelling and capitalization.

SingleServo

```
#include <Servo.h>
Servo myServo;

void setup() {
  myServo.attach(9);
}

void loop() {
  myServo.write(0);
  delay(2000);
  myServo.write(180);
  delay(2000);
}
```

When you're done typing the sketch:

1. Save the program for future use (name it *SingleServo.ino,* or something similar so you'll remember it).
2. Select your model of Arduino development board in the Board list (*Tools | Board*).
3. Click the Verify button to compile the sketch. If there are any typographical or syntax errors, the IDE will flag them in the status window at the bottom. Double-check, fix any errors you find, and try again.
4. Connect the USB cable between the Arduino and your PC. The On indicator light on the Arduino should light up.
5. Verify that the proper communications port is selected by choosing *Tools | Serial Port.* Select the desired port if there's not already a check beside it.
6. Click the Upload button to transfer the sketch to your Arduino. When finished the status window will indicate that uploading has completed.

After a few seconds' delay, the servo motor should spin in one direction for two seconds, then stop and spin in the other direction for two seconds. This will repeat until you unplug the Arduino.

ANALYZING THE ONE SERVO SKETCH

The one servo sketch is short and to the point. Here's how it works.

Including the Servo Library

```
#include <Servo.h>
```

The first line is a reference to one of the Arduino libraries. A *library* adds more features to the Arduino than what's in the built-in command set. As its name suggests, the *Servo* library provides additional programming statements tailor-made just for operating servos.

Note that *Servo* is a "built-in" library, in that it comes with the Arduino IDE software. You can add other libraries that extend the Arduino even further. We won't get to this topic in this chapter, but it's covered and demonstrated in many of the chapters that follow.

Something didn't work right? Make sure you have version 1.0 of the Arduino IDE installed on your computer. This ensures you're using a compatible version of the *Servo* library.

In many cases, libraries are made to greatly simplify some complex programming task. In other cases, they allow for reuse of code, so that your sketches are as small and compact as possible. The *Servo* library is a good example of both: It decomplicates a complicated internal mechanism used to operate a servo motor, and it provides a means to control not one motor, but up to 12 (that's for the basic Arduino Uno; you can control up to 48 servos with the Arduino Mega 2560).

Next is the line

```
Servo myServo;
```

Many libraries, including *Servo,* use a programming concept known as *classes.* In programming parlance a class is a package consisting of data, along with various mechanisms that operate on that data. You don't work with the class directly; instead, you create an *object* using that class. In the previous line,

```
Servo
```

is the class, which is contained in the *Servo* library you reference at the top of the sketch. And

```
myServo
```

is the name of the object you created using the *Servo* class. Using a simple real-world analogy, suppose Cat is a class and felix is a specific individual ("object") belonging to that class. In Arduino-speak, you'd identify felix the cat with the line:

```
Cat felix;
```

The name of the object is up to you, but it must follow the standard naming practice observed for user-defined variables and functions that you learned about in Chapter 2, "Arduino Up and Running."

In quick review: The name can only contain letters and numbers or the underscore character. It must start with a letter. You can mix and match letter case, but you must be sure to always follow the same capitalization. The name *myServo* is treated as a different object from one called *MyServo.*

Finally, remember that object names can't be the same as any programming statements already defined for the Arduino.

In programming lingo the line *Servo myServo* creates, *constructs,* or *instantiates* a servo *object.* I mention these phrases now, as you may encounter them as you study various Arduino sketches. The functionality of this object is defined in the *Servo.h* library and its accompanying *Servo.cpp* programming code file.

If you're curious what's inside these files you can look at them using any text editor. The libraries that come with the Arduino are contained in the libraries folder under the main Arduino program directory. Avoid making any changes in them until you're fully familiar with Arduino programming.

Setting Up the Servo in Setup

```
void setup() {
  myServo.attach(9);
}
```

Next is the mandatory *setup* function, containing a single statement. Notice the unusual syntax:

- *myServo* is the name of the servo object that was defined earlier.
- *attach* is a *method* that you can use with the myServo object. Methods are actions that you use to control the behavior of objects. In this case, *attach* tells the Arduino that you have physically connected the servo to digital pin D9 and you want to activate it. A period separates the object name and method—*myServo.attach.*

In the case of *attach*, the method tells the Arduino which of its I/O pins is connected to the servo. The signal wire of the servo is attached to digital pin 9 (D9), so the value *9* is used as the *argument* for the *attach* method.

Running the Servo Back and Forth

```
void loop() {
  myServo.write(0);
  delay(2000);
  myServo.write(180);
  delay(2000);
}
```

The *loop* function contains the part of the sketch that is repeated over and over again, until you download a new program or remove power from the Arduino. The *loop* function contains four lines:

myServo.write(0) is another method using the myServo object. The *myServo.write(0)* method instructs the servo to move all the way in one direction. When using a modified servo this statement causes the motor to continually rotate in one direction.

delay(2000) tells the Arduino to wait the specified period of time, in this case 2000 milliseconds, or two seconds.

The two statements are repeated again, this time with *myServo.write(180),* to make the servo go the other direction (see Figure 4-2).

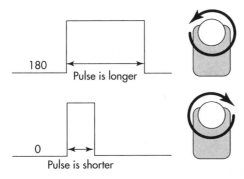

180
Pulse is longer

0
Pulse is shorter

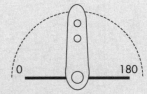

Figure 4-2 The servo *write* method accepts values from 0 to 180, which correspond to increasing pulse duration. You can change the direction of a modified R/C servo by using 0 or 180.

The *write* method expects a value of 0 to 180. These numbers represent angles in degrees. Ordinarily, these angles refer to the −90 and +90 sweep of the typical unmodified servo (0–180 is used to avoid negative numbers).

Not all servos turn the same direction when given the values 0 and 180. Depending on the make and model of the servo, the value 0 may turn the motor counterclockwise, or it may turn it clockwise. The direction of rotation isn't usually critical, as you can always compensate for the differences among servos in your Arduino sketches. Just transpose the 0's and 180's.

Wiring for Two Servos

The Arduino natively supports up to 12 servos (48 with the Arduino Mega 2560). From a programming standpoint, adding more motors is little more than making additional *Servo* objects and controlling each one with its own *write* method.

Connecting an R/C servo directly to the Arduino board is acceptable when using just one motor. But it's not advisable with two or more servos, because when they're all running they'll draw more current than what the voltage regulator on the Arduino can handle (and, if you're using the USB cable to provide juice, more than what your PC can deliver). So instead, you'll want to use a split power supply, which uses a separate battery pack for the servo motors.

The first thing to do is set up the twin breadboards on the top deck of the Teachbot as shown in Figure 4-3. For each board separate one of the power rails, and position as shown in the diagram. Use the tongue-and-groove to lock the four pieces together. As demonstrated in Chapter 3 use hook-and-loop (Velcro) to secure the breadboards to the top deck, surrounding the Arduino board.

The solderless breadboards demonstrated here and used with the prototype Teachbot use a tongue-and-groove to mate the upper and lower power rails with the main body of the breadboard. Separate these parts by sliding them off.

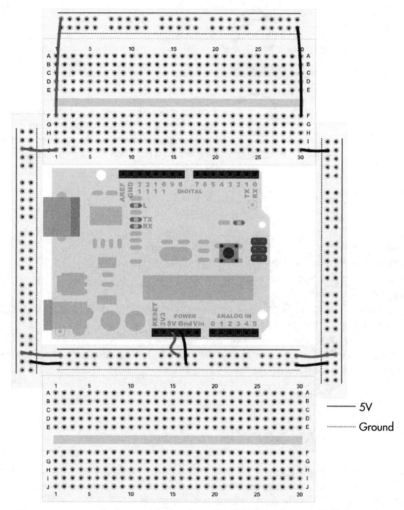

5V

Ground

Figure 4-3 Interlocking solderless breadboard layout for the Teachbot Servo robot. Use hook-and-loop (Velcro) to secure the large breadboard pieces into place.

Finish the breadboards by wiring the red (+) and black (–) power. Be sure not to cross up this wiring or it could cause a short circuit. (The Arduino will handle this by shutting down, but you don't want the condition to last for very long.)

The wiring schematic for a split supply is shown in Figure 4-4; the same plan is depicted in breadboard view in Figures 4-5a and 4-5b. These illustrate how to attach two servo motors to the Arduino.

- The Arduino is powered by its own supply, which can be the USB cable from your computer, a plug-in wall transformer delivering 7–12 volts, or a suitable battery, like a 9-volt cell.
- The servos are powered by their own supply. I like to use a set of AA or AAA cells for servos on a mobile robot. You can opt for nonrechargeable or rechargeable cells.

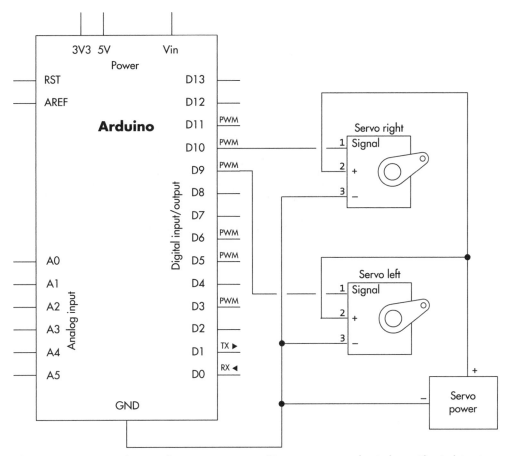

Figure 4-4 Wiring diagram for connecting two R/C servo motors to the Arduino. The Arduino is powered from its own source (not shown here; it can be USB or its power plug). Note the separate power for the servos (4.8 to 6 volts from a set of batteries) and the common ground connection. This is very important.

Remember that nonrechargeable batteries deliver a nominal (average) voltage of 1.5 volts per cell; rechargeable, 1.2 volts per cell. Put the batteries in a suitable multicell holder depending on the voltage you want to supply to the servos.

Number of Cells	Nonrechargeable	Rechargeable
4	6 volts	4.8 volts
5*	7.5 volts	6 volts
6	9 volts**	7.2 volts

*Five-cell battery holders are not common, but are available special order from some electronics distributors. You can make your own by combining a 2- and 3-cell holder, or by using a 6-cell holder with one of the "cell pockets" shorted through. Both techniques are shown in Figure 4-6.

**Powering the motor with 9 volts will damage most servos. The voltage is shown here only for comparison purposes.

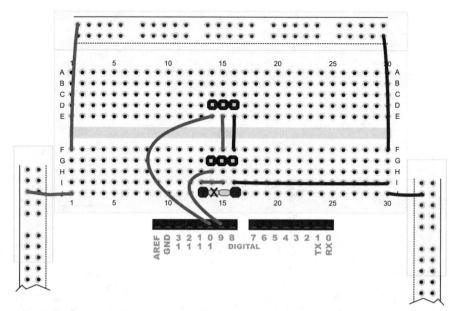

Figure 4-5a Wiring connections on the upper solderless breadboard for the servo connections and servo battery. Important! Be sure to observe correct polarity for the servo wiring, or damage to the servos will result.

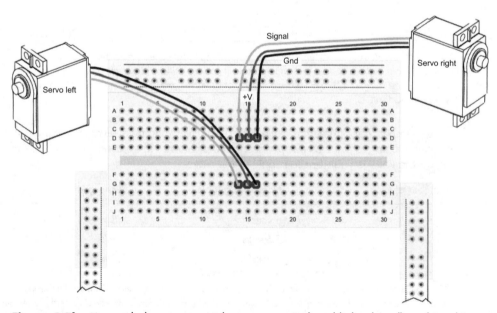

Figure 4-5b How and where to connect the two servos on the solderless breadboard (combine this wiring with what you see in Figure 4-5a)

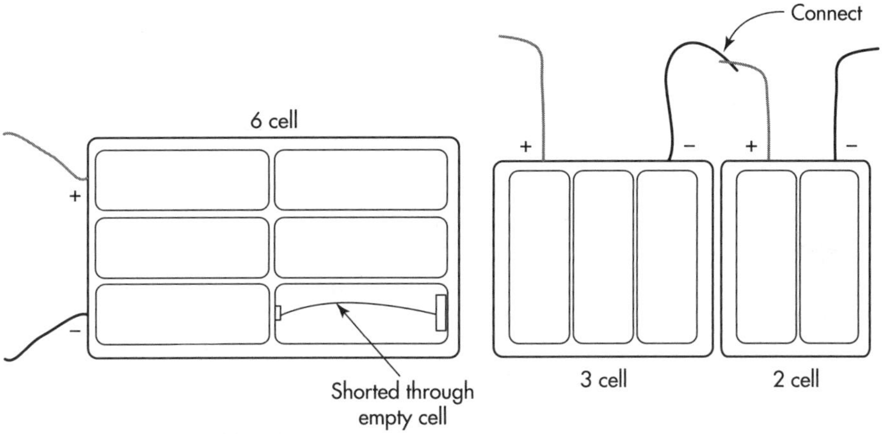

Connect

6 cell

+

−

Shorted through
empty cell

+   −   +   −

3 cell   2 cell

Figure 4-6   You can construct a 5-cell battery holder for providing 6 volts from
rechargeable batteries by either shorting through an empty cell pocket of a 6-cell holder,
or combining a 2- and 3-cell holder, as shown.

I've specified a 4-cell holder for the Teachbot, and this is what's shown in the construction
details in Chapter 2. You can use either rechargeable or nonrechargeable cells. When using
rechargeable batteries, opt for the nickel-metal hydride (NiMH) type; they're more environ-
mentally friendly. (Be sure to charge them in a recharger made for NiMH batteries, or else
they will be damaged.) You'll want alkaline cells if using nonrechargeable batteries. Ordinary
zinc cells, even the "heavy duty" type, will wear down very quickly.

Most servos are engineered for a 4.8 to 6 volt supply, though on many you can go as high as
7.2 volts (on some, even up to 8.4 volts) without damaging the motor. But check the
manufacturer's data sheet first!

    The higher the voltage, the faster the motor turns, and the more torque—ability to provide
mechanical power—it has. For servos that have been modified to spin continuously, the
higher voltage means the robot travels a bit faster and can carry a heavier load.

    These are compelling reasons to operate your robot's servos at the highest voltage you can
reasonably provide, but avoid going beyond what the manufacturer recommends for its
product. Otherwise, you may very well end up burning out the motor.

Which type of battery cell should you choose—rechargeable or nonrechargeable? I prefer
rechargeable batteries, as they're overall cheaper, and not as wasteful for the environment.
While the voltage output of rechargeable cells is a little lower (and therefore the robot won't
move quite as fast) one set of batteries will last months, even years.

BREADBOARD CONNECTIONS

The twin motors on the Teachbot are electrically connected to the Arduino by way of a solderless breadboard. This technique is used because the Arduino Uno board (and similar models)
lacks direct connections for attaching to servo motors. The breadboard is wired as shown in
Figures 4-5a and 4-5b to provide a conduit for both power and signal.

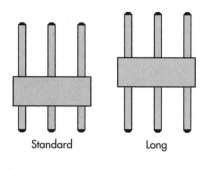

Standard   Long

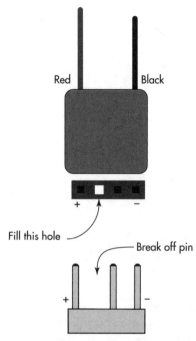

Red   Black

Fill this hole

Break off pin

+   −

Figure 4-7   How to create a polarized power plug for the servo battery pack. The broken pin, and corresponding filled hole in the connector, help prevent you from plugging the pack in backwards.

R/C servos come with a three-pin female header, which can plug into the solderless breadboard using 0.100" double-sided (long) male header pins. These header pins usually come in long strips; break off two sets of three pins and one set of pins for the AA battery connection.

You want the version of male header pins that are "double-sided"—they're long on both sides. If you use standard header pins, the length of pins on one side is shorter. These don't make good contact when used with solderless breadboard designs.

My reference design for the Teachbot uses a AA battery holder with a four-pin female connector. The + and − leads are on the two outside positions of the connector. I've broken off the pin right next to the + connection of the male header (see Figure 4-7), then filled in the corresponding hole with a small dollop of fast-setting goo from an ordinary glue gun. This prevents the connector from being reversed when plugging in.

When wiring the solderless breadboard, be especially careful not to mix positive and negative leads to the servo. *Reversing the power leads may permanently damage the servo.*

 The Teachbot uses separate battery supplies for the Arduino and its two servos. In order for everything to function properly, the ground connections for Arduino power and servo power must be connected together. This is shown in both the schematic and pictorial circuit views. If you don't do this, the servos may not function at all, or they may behave erratically.

Make sure to also properly orient the connectors for the servos when you plug them into the board. Servo power leads are color-coded, but as noted earlier in the chapter the colors aren't universal. Figure 4-8 shows the Teachbot with solderless breadboards attached and wired with servos.

(And yes, I'm using an older Arduino Diecimila as a photo model. All the Teachbot Servo examples were prepared using an Arduino Uno, but they'll also work with the Diecimila and Duemilanove boards, and very likely many other Arduino-compatible boards.)

RUNNING THE SERVO TEST SKETCH

With the Teachbot constructed and the breadboard wired, you're ready to test the robot and put it through its paces. Refer to the *TeachbotServo_2MotorDemo* sketch.

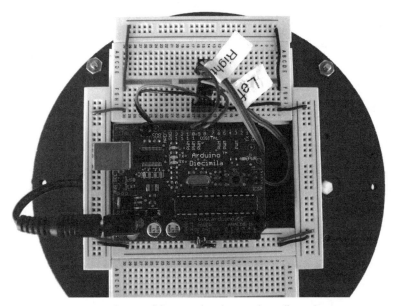

Figure 4-8   Pictorial view of the completed upper breadboard area, servos, and servo power. I've put labels on the servo wires for easier identification.

TeachbotServo_2MotorDemo

```
/*
 Teachbot ServoTest
 Tests servos of robot by moving them in different directions
 Requires Arduino IDE version 1.0 or later

*/

#include <Servo.h>
Servo servoLeft;          // Define left servo
Servo servoRight;         // Define right servo

void setup() {
  servoLeft.attach(9);    // Set left servo to digital pin 9
  servoRight.attach(10);  // Set right servo to digital pin 10
}

void loop() {             // Loop through motion tests
  forward();              // Example: move forward
  delay(2000);            // Wait 2000 milliseconds (2 seconds)
  reverse();
  delay(2000);
  turnRight();
  delay(2000);
  turnLeft();
  delay(2000);
  stopRobot();
  delay(2000);
}
```

```
// Motion routines for forward, reverse, turns, and stop
void forward() {
  servoLeft.write(0);
  servoRight.write(180);
}

void reverse() {
  servoLeft.write(180);
  servoRight.write(0);
}

void turnRight() {
  servoLeft.write(180);
  servoRight.write(180);
}

void turnLeft() {
  servoLeft.write(0);
  servoRight.write(0);
}

void stopRobot() {
  servoLeft.write(90);
  servoRight.write(90);
}
```

Start the Arduino IDE, connect a USB cable between your computer and the Arduino, and type the program as shown. You can omit the comments if you'd like. When done, verify (compile) the sketch, and look for any syntax errors. If there are none, download the sketch to your Arduino.

Once downloaded, put a small book under your Teachbot to lift its wheels off the ground. Disconnect the USB cable, and in this order, plug the AA battery connector into the breadboard, then plug in the 9 volt power to the Arduino power jack. (If you are using an earlier Arduino Diecimila be sure to switch over the power selection jumper from USB to EXTernal.)

If everything is connected properly, the servo motors should go through a test pattern. Assuming the motors are working as they should, depress the Reset switch on the Arduino board, and place the Teachbot on the ground. Release the switch and the robot should perform its self-test maneuvers.

If the motors aren't moving, double-check your wiring, making sure the servo connectors are properly oriented. They won't work if the connectors are reversed on the breadboard.

A CLOSER LOOK AT THE TEST SKETCH

The two-servo demo sketch is straightforward. As with the single-servo example, the sketch begins with an *include* statement to the Servo.h library header file, which is provided with the Arduino IDE installation. This file, and its corresponding C-language program, provide all the actual coding to make the servos function.

Next comes two statements that create (instantiate) two *Servo* objects for use in the remainder of the sketch. Each object represents a physical servo attached to the Arduino. Methods of these objects include things like specifying which digital pin is used to connect from the Arduino to the servo and the position of the servo. Note I've given the two *Servo* objects descriptive names: *servoLeft* and *servoRight*. It's easier to keep track of things this way.

In the *setup* function the *servoLeft* and *servoRight* objects are "wired" to their respective pins on the Arduino; in this case, pin 9 for *servoLeft* and pin 10 for *servoRight*.

Now comes the main body of the program, provided in the *loop* function. It contains a series of user-defined functions for *forward, backward,* and so on, plus a delay of 2000 milliseconds (two seconds) between each function. You can see that the robot repeats the same demonstration steps over and over:

- Goes forward for two seconds
- Reverses for two seconds
- Turns right for two seconds
- Turns left for two seconds
- Stops for two seconds

And finally, each user-defined function specifies the motion to apply to the servos. With the *Servo* object, servos are commanded to move one direction or another by (among other ways) specifying an angle of between 0 and 180. The servo then moves to that angle in response.

When using servos that have been modified for continuous rotation, 0 makes the servo rotate one direction, 180 makes the servo rotate in the opposite direction, and 90 makes it stop.

Controlling the Teachbot Servo Using Wired Control

I like to demonstrate the control and operation of two-wheeled robots using manual remote control, as it shows how the motion of the bot is ruled by the direction of its motors.

Figure 4-9 shows a simple and low-cost method of wiring a remote control box to the Teachbot. You can use this remote to try out different motions, and practice how to move the robot through a room.

The project uses a standard Nintendo Entertainment System (NES) gamepad controller, along with a special controller adapter. The adapter plugs into the solderless breadboard on the Teachbot. See Figure 4-10 for a diagram for wiring the controller to the Arduino.

You need the basic and old-fashioned NES gamepad controller, not any of the extended Nintendo controllers such as Zapper or PowerPad. If you don't already have a compatible NES controller lying around the house, you can often find these new or used on the Web.

I'm using a replacement controller available from Parallax. The breadboard-compatible controller adapter is also from Parallax.

See *TeachbotServo_NES.ino* for the Teachbot gamepad controller sketch.

TeachbotServo_NES

```
/*
   NES gamepad button values
   right  =  11111110    bit 0
   left   =  11111101    bit 1
```

Figure 4-9   Manually control the motors on your Teachbot Servo using a Nintendo Entertainment System gamepad and adapter (both available from Parallax). The controller accepts up to two pads; plug the gamepad into the socket on the left.

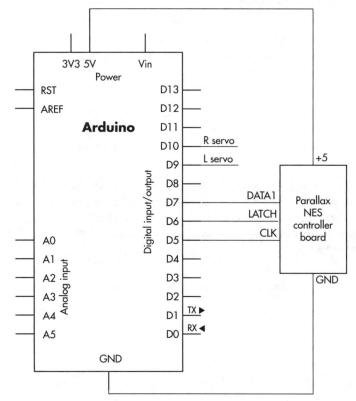

Figure 4-10   Wiring diagram for the NES gamepad, using the Parallax controller

```
    down   =  11111011   bit 2
    up     =  11110111   bit 3
    start  =  11101111   bit 4
    select =  11011111   bit 5
    b      =  10111111   bit 6
    a      =  01111111   bit 7
*/

#include <Servo.h>

#define clock     5
#define latch     6
#define dataIn    7
#define LServoPin 9
#define RServoPin 10

byte nes = 0;
Servo servoL;
Servo servoR;

void setup() {
  servoL.attach(LServoPin);
  servoR.attach(RServoPin);
  pinMode(clock, OUTPUT);
  pinMode(latch, OUTPUT);
  pinMode(dataIn, INPUT);
  digitalWrite(latch, LOW);
  digitalWrite(clock, LOW);
}

void loop() {
  nes = nesRead();
  if (bitRead(nes, 0) == 0) {
    servoL.write(180);
    servoR.write(180);
  }

  if (bitRead(nes, 1) == 0) {
    servoL.write(0);
    servoR.write(0);
  }

  if (bitRead(nes, 2) == 0) {
    servoL.write(180);
    servoR.write(0);
  }

  if (bitRead(nes, 3) == 0) {
    servoL.write(0);
    servoR.write(180);
  }

  if(nes == 255) {
    servoL.write(90);
    servoR.write(90);
  }

  delay(200);
}
```

```
byte nesRead() {
  byte value = 0;
  digitalWrite(latch, HIGH);
  delayMicroseconds(5);
  digitalWrite(latch, LOW);
  for (int i=0; i<8; i++) {
    digitalWrite(clock, LOW);
    value |= digitalRead(dataIn) << (7 - i);
    digitalWrite(clock, HIGH);
  }
  return(value);
}
```

Only the Up/Down/Left/Right direction buttons of the controller are used in the sketch; the A, B, Select, and Start buttons are unassigned. The direction buttons activate the servos to move forward or backward, or turn right or left. Release the direction buttons and the robot stops.

Here's how the sketch works:

Servo objects, constants, and global variables are defined at the top of the sketch. Defined constants are like those made with the *const* keyword, but they use a different syntax—note, especially, there is no equals sign between the constant name and its value, nor is there a semicolon at the end of the line. A *global* variable (it's visible to all parts of the sketch), named *nes,* is defined here as well, along with two *Servo* objects.

The *setup* function attaches the two servos and prepares the three pins connected to the NES controller. These pins, labeled for our convenience as *clock, latch,* and *dataIn,* are used to transfer data from the controller to the Arduino. The *clock* and *latch* pins are made OUTPUTs; the *dataIn* pin is an INPUT. Before leaving the *setup* function the *clock* and *latch* pins are made LOW (0 volts).

The *loop* function calls the user-defined *nesRead* function (see later), and stores its 8-bit (byte) value in the *nes* variable. The individual bits of this value are checked to see if a corresponding button is pressed. We're only interested in the left, right, up, and down buttons, and if any of these are pressed, the servos are moved accordingly. Note the if test for *nes == 255.* This checks to see if the value is 255 (all 1's, or 11111111). That means no button is pressed. In that case the robot stops.

The *nesRead* user-defined function retrieves data from the NES controller. It does this using a *bit-banging* technique of sending a clock pulse, then fetching whatever single bit of data the controller returns. The process of "clocking in" the data is repeated a total of eight times, for all eight bits.

The Arduino IDE supports a special function called *shiftIn* that does a similar job as the *nesRead* function. I've opted for a custom job here, as the *shiftIn* function behaves differently depending on the version of the IDE you're using.

The NES controller works in such a way that the first bit is already available before clocking has begun. Depending on the specific version of the Arduino IDE, when using the *shiftIn* function the data isn't fetched until *after* each clock cycle. That means the first bit will be lost (it's the bit for the A button), and all the remaining bits shifted over by one position. Not ideal.

More About the Servo Object

You've read that *Servo* is a class library you can import into your sketches, and this class allows you to create one or more *Servo* objects; each object controls a physical servo motor attached to the Arduino. For each object there are numerous methods—special commands specific to the *Servo* object—for operating your motors.

You've already used *attach* and *write;* but for the purposes of review here are the principal *Servo* object methods you should know about:

attach connects a servo object to a specific pin of the Arduino. You can use any pin, including the analog pins A0 through A5.

detach removes the servo object, effectively disabling the servo and removing its power.

write specifies where to position the servo. The method accepts several forms of values. A value of 0 to 180 denotes degrees; this positions the shaft of the motor to a corresponding angle. (When used with modified servos, 0 and 180 make the motor turn one direction or the other; 90 makes the motor stop.)

writeMicroseconds specifically indicates you wish to use microseconds to control the servo position. Use this method when you want to explicitly control the precise pulse duration provided to the servo.

attached checks if a servo motor is currently attached to an Arduino pin. You can use this method to determine if a servo needs to be reattached.

read returns the last specified position of the servo, in degrees. This method has limited use with modified servos, but can come in handy with nonmodified motors. Note that the *Servo.read* method doesn't actually report the physical position of the servo, just the value of the last *write* or *writeMicroseconds* command you sent it.

FYI In addition to what I cover in this book, I recommend you check out the documentation for the *Servo* library on the *www.arduino.cc* Web site. You'll find additional examples and a complete list of all *Servo* object methods.

You can send microsecond values from the *write* method as well. When using *write,* values between 0 and 180 are construed as measurements in degrees. These map to a built-in span of between 544 and 2400 microseconds.

The typical servo turns in a 90 to 120 degree arc using a 1000 to 2000 μs span. Most motors accept values below and above these timings, as shown in Figure 4-11. When using a nonmodified motor you must be careful when applying pulses outside the normal range; otherwise, the motor may hit against an internal stop, and this could damage it. This problem doesn't exist with modified servos, which have had their stops removed.

Values above 180 are treated by the *write* method as microseconds, rather than degrees. As with *writeMicroseconds,* you can use these values to position the servo by generating pulses of the specified duration.

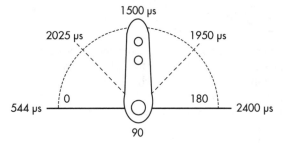

Figure 4-11 The servo *write* method accepts either angles in degrees (0 to 180) or direct microsecond durations. Not all servos provide full 180-degree rotation given the extremes shown here. If your servo "chatters" or buzzes loudly at either extreme, it may be hitting an internal stop. Adjust the values accordingly to avoid this condition.

ATTACHING AND DETACHING SERVOS

Refer back to the *TeachbotServo_2MotorDemo* sketch, and note the *stopRobot* user-defined function. It demonstrates stopping the two servos by sending it commands to position the motors at their neutral 90-degree position (equates to 1500 µs). For servos that have been modified for continuous rotation, this is meant to stop the motors.

Unfortunately, the motors may not fully stop with 1500 µs pulses. They may "creep" slowly one direction or the other. This is caused by slight variations in the circuitry within the servo, most often from changes in temperature. As the electronics inside the servo get warm, the 1500 µs centerpoint wavers. Sending the servo pulses of this length no longer fully stops it.

Some servos that come premodified for continuous rotation have a hole in the side or top to allow you to tweak the potentiometer inside the motor. This lets you readjust the centerpoint to make the motor once again stop when it receives 1500 µs pulses. Trouble is, under different operating temperatures you may be tweaking the innards of the motor more or less constantly.

A better method is to "detach" the servo. This causes the Arduino to stop sending it pulses. Without pulses the motor stops (this is normally true, but see the Tip that follows).

```
void stopRobot() {
  servoLeft.detach();
  servoRight.detach();
}
```

Once detached using the *detach* method, you can always reattach it. Each of the motion routines requires first reattaching the two motors, then sending position instructions using the *write* method:

```
void forward() {
  servoLeft.attach(9);
  servoRight.attach(10);
  servoLeft.write(0);
  servoRight.write(180);
}
```

Detaching a servo to stop it applies to analog servos, but not all digital servos. With many digital servos, removing the pulses has no effect: Digital servos will continue to hold position even when the pulses go missing. This is one of the reasons digital servos aren't always a good option for use as continuous rotation motors.

A somewhat more elegant solution to reattaching the servos is provided in the *Teachbot-Servo_DetachDemo* sketch. Here, each of the motion routines calls a user-defined function, *reAttach,* that checks if the servos are not attached. If they are not attached, the code reattaches them.

Note the way the if test expression works: *!servoLeft.attached().* The *attached* method returns a *boolean* true or false value. The exclamation point (called a *bang* in programmer-speak) negates the value returned by the method. When reading the statement you'd say aloud "if not servoLeft dot attached." (Or you could say it to yourself, but talking to yourself makes people think you're an eccentric genius.)

Another, slightly longer, way of writing the if expression would be:

```
if(servoLeft.attached() == false)
```

TeachbotServo_DetachDemo

```
#include <Servo.h>
Servo servoLeft;          // Define left servo
Servo servoRight;         // Define right servo

void setup() {
   servoLeft.attach(9);   // Set left servo to digital pin 9
   servoRight.attach(10); // Set right servo to digital pin 10
}

void loop() {
   forward();
   delay(2000);
   reverse();
   delay(2000);
   turnRight();
   delay(2000);
   turnLeft();
   delay(2000);
   stopRobot();
   delay(2000);
}

void forward() {
   reAttach();
   servoLeft.write(0);
   servoRight.write(180);
}

void reverse() {
   reAttach();
   servoLeft.write(180);
   servoRight.write(0);
}

void turnRight() {
   reAttach();
   servoLeft.write(180);
   servoRight.write(180);
}

void turnLeft() {
   reAttach();
   servoLeft.write(0);
   servoRight.write(0);
}
```

```
void stopRobot() {
  servoLeft.detach();
  servoRight.detach();
}

void reAttach() {
  if(!servoLeft.attached())
    servoLeft.attach(9);
  if(!servoRight.attached())
    servoRight.attach(10);
}
```

STARTING WITH A SPECIFIC POSITION VALUE

When you create a new *Servo* object, its position is automatically given a default of 90 (neutral or stopped in the case of a servo modified for continuous rotation).

You can instead write a position to the servo *before* calling the *attach* method—attaching the servo is what gives it power. By setting a position first, then attaching the servo, you can have it start at a position other than 90 degrees. For example:

```
void setup() {
  servoLeft.write(180);     // Start at 180 degrees instead of 90
  servoLeft.attach(9);
}
```

"Prewriting" a starting value before powering the motor is of most use with unmodified servos. One use is to set unique starting positions for all the servos in your robotic arm, so that when power is applied the thing doesn't flail in the air like the Robot in the old *Lost in Space* TV show. (In fact, this exact procedure is demonstrated in Chapter 13, "Robby Armstrong"—the anti-flailing part, not the wacky *Lost in Space* robot.)

Still, this technique also has its place for modified servos. For example, you may find that your servos fully stop when their neutral position is set to 1550 μs, rather than the usual 1500 μs. You can start up your robot and stop the motors by prewriting this revised neutral position, then attaching the servo, like so:

```
servoLeft.write(1550);  // Or you can use writeMicroseconds
servoLeft.attach(9);
```

SETTING SPEED OF ROTATION

You can control the speed of servos that have been modified for continuous rotation. The technique involves setting a pulse duration close to the 1500 μs neutral position. The closer the pulses are to this timing, the slower the motor will turn.

The exact speeds you get at different pulse durations depend on the motor. Figure 4-12 demonstrates a typical speed curve from 1000 to 2000 μs. Notice that the biggest speed

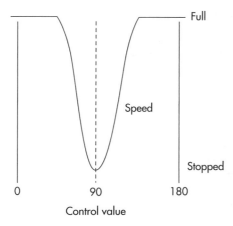

Figure 4-12 Control the speed of a continuous rotation servo by providing signals near the (90 degree, or 1500 μs) centerpoint. On most servos, speed adjustment is not linear across the 0–180 range.

difference is right around the 1500 μs neutral point. As you move further away from the neutral center, the speed difference is less apparent.

You can use the behavior to make the robot slow down. For going straight ahead, you can reduce the speed by altering the values by 10 or 15 points, like this:

```
void forward() {
  servoLeft.write(80);
  servoRight.write(100);
}
```

Speed control is most useful in turns; the slower the turn, the more accurate it can be. These values are 20 points from 90 (neutral):

```
void turnRight() {
  servoLeft.write(110);
  servoRight.write(110);
}
```

You may notice that the two servo motors on your Teachbot don't exhibit the same reduced speed, causing the robot to veer to the left or right (this isn't a problem in turns, but can be when going straight forward or backward). You can compensate for any speed difference by carefully adjusting the values used in the *Servo.write* methods for both motors. Try slightly higher or lower values.

Specifying servo speeds in microsecond units offers better resolution—you have at most only 180 steps when using the 0–180 angle scale with the *write* method. By switching to microseconds instead you can try all the in-between values.

Note, though, that most servos don't respond to changes as little as four or six microseconds' difference. This is due to *dead band,* an inherent feature of most servos where very small changes in pulse duration don't have any noticeable effect on the output of the servo. You might try changing the duration in five-microsecond increments.

REFERRING TO I/O PINS USING A VARIABLE

You can make your sketches with servos easier to adapt and modify for other applications by using variables for the I/O pins used to connect to each motor.

Variables are defined in an Arduino sketch by first indicating the type of variable you want to use, the name of the variable, and then its value. For example:

```
int sLeft = 9;
```

tells the Arduino you're creating an *int* variable type (16 bits; see the discussion in Chapter 2 for more details). The *sLeft* is the name you're giving to this particular variable, and 9 is the pin number. Notice the = (equals) sign and semicolon at the end of the line. These are necessary.

Here's the first part of the *TeachbotServo_2MotorDemo* sketch, modified to use variables to hold the Arduino pin numbers each servo is attached to:

```
#include <Servo.h>
Servo servoLeft;
Servo servoRight;
int sLeft = 9;              // Specifies left servo on D9
int sRight = 10;           // Specifies right servo on D10

void setup() {
  servoLeft.attach(sLeft);
  servoRight.attach(sRight);
}
// . . . and so on
```

Recall from Chapter 2 that variables where the content never changes are referred to as *constants*. The values in *sLeft* and *sRight* won't change throughout the running of the sketch, so to save memory space you can convert these variables to constants. As a constant, the value is set in the compiled sketch, and doesn't take up memory space in the Arduino's microcontroller.

It's easy to do: Just add the keyword *const* (for constant) in front of both variable assignments, as in:

```
const int sLeft = 9;
const int sRight = 10;
```

There are other ways of accomplishing the same thing in an Arduino sketch (specifically, you can use the *#define* keyword in most cases), but *const* is more straightforward, and is the method the Keepers of the Arduino recommend.

SERVO MOVEMENT WITHOUT DELAYS

By now you've surely noticed the frequent use of *delay* statements in the foregoing servo demonstrations. *Delay* causes the Arduino to wait for a specified period of time before continuing. Without the delays the Arduino would zoom through all the programming statements before the motors have a chance to react.

Delays with long periods of waiting (anything more than about a half-second) are acceptable for demonstrations, and you can still occasionally use them in more elaborate programs. But there are other methods you can use to time the action of the servo motors in your robot. Several methods are introduced and discussed in the chapters to follow.

Limitations of Modified Servos

Before closing out this chapter let's take a brief moment to ponder some of the limitations of using modified servos in a robot. Here are some prominent points you'll want to keep in mind:

- The average servo is not engineered for lots and lots of continual use. The mechanics of the servo are likely to wear out after perhaps as little as 25 hours (that's total elapsed time), depending on the amount of load on the servos. Models with metal gears and/or brass bushing or ball bearings will last longer.
- Standard-sized servos are not particularly strong in comparison to most other DC motors with gear heads. Don't expect a servo to move a five- or ten-pound robot. If your robot is heavy, consider using either larger, higher-output servos (such as 1/4-scale or sail winch), or better yet, DC motors with built-in gear heads.
- Modifying a servo voids its warranty. You'll want to test the servo before you modify it to ensure that it works.

Chapter 5

Programming the Teachbot: Seeing It React

Robots need information about the world around them, or they just stumble around looking stupid. Just like us humans, a robot uses senses to know when it's run into something; when it's light or dark; when it's too hot or too cold; when it's about to fall over; when it's found the way to the cheese at the center of a maze.

Senses require sensors. And in the practice of robotics, the basic senses make do with the most basic of sensors—mechanical switches for detecting contact with objects, and photosensitive resistors and transistors for detecting the presence (or absence) of light. A robot can perform a remarkable amount of work with just the sense of touch and the gift of simple sight.

In this chapter you'll learn about interfacing switches and photosensors to the Teachbot Servo, along with how to use the information these sensors provide to interactively command a robot's motors. These are the fundamental building blocks of most any autonomous robot. Once you learn how to use these sensors to do your bidding, you can apply them in dozens of ways, for all kinds of robotic chores.

 Be sure to visit the ARB Support Site (see Appendix A) for additional code examples and variations. All sketches from this book are available for ready download.

 FYI Brand new to Arduino programming? Then be sure to first read through Chapter 2, "Arduino Up and Running." And, of course, refer to the official Arduino Language reference pages at *www.arduino.cc*.

Getting in Touch with Your Robot

Sensors—whether in humans or in robots—are designed to produce a reaction. What that reaction is depends on the nature of the sensation. Type and quantity matter. We interpret the feeling of a soft summer breeze as a good sensation. But increase the amount of air pressure to hurricane force, and decrease the temperature to something below freezing, and now the same senses produce a highly negative reaction.

Touch, also called *tactile feedback,* is a primitive reactive sense. The robot determines its environment by making physical contact; this contact is registered through a variety of touch sensors. What happens when contact is made is entirely determined by the programming you apply within your robot.

Most often, a collision with an object is a cause for alarm. So the reaction of the robot is to stop what it's doing and back away from the condition. But in other cases, contact can mean your robot has found its home base, or that it's located an enemy bot and is about to pound the living batteries out of it.

The lowly mechanical switch is the most common, and most simple, form of tactile (touch) feedback mechanism. Just about any momentary, spring-loaded switch will do. When the robot makes contact, the switch closes, completing a circuit.

I like to use *leaf* (or *lever*) switches (see Figure 5-1) because they function a lot like a cat's whiskers. These things are sometimes referred to as a Microswitch, after a popular brand name, but I'll call them leaf switches to avoid confusion. Regardless of make or model, most are easy to mount, and come with plastic or metal strips of different lengths that enhance the sensitivity of the switch.

You can enlarge the contact area of the leaf by gluing or soldering bigger pieces of plastic or metal to it. For example, you can cut up some stiff music wire (available at hardware stores) or a cheap wire clothes hanger, and bend it to some fancy shape. Solder the end(s) to the leaf. Or you can use thin pieces of wood, plastic, or metal. Just be sure the weight of the extension doesn't accidentally activate the switch.

The switch may be directly connected to a motor, or more commonly, it may be connected to a microcontroller. A typical wiring diagram for the switch is shown in Figure 5-2.

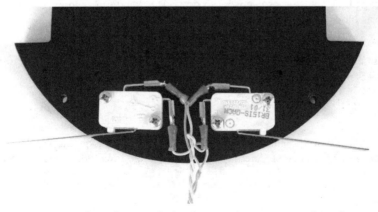

Figure 5-1   Leaf switches are ideal as bumper detectors on a robot. Place two at the front and the robot can detect contact with obstacles to the left or right.

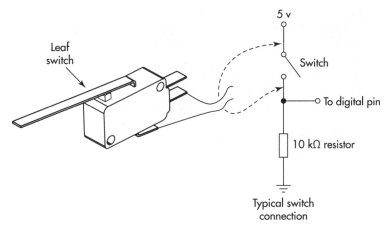

Figure 5-2 The leaf switch acts as a kind of cat's whiskers. It's connected to a digital I/O pin using a pull-down resistor.

The 10 kΩ *pull-down resistor* is there to provide a consistent digital LOW (0 volts) output for the switch when there is no contact. When contact is made, the switch closes, and the output of the switch goes HIGH—usually 5 volts, as shown in the figure.

What's the purpose of a pull-down resistor? The resistor literally "pulls down" the voltage of the Arduino's input pin to 0 volts (LOW), preventing the pin from "floating" and giving erroneous results. The resistor allows the contacts within the switch to push the input to a 5 volt (HIGH) state without causing a short circuit.

A pull-down resistor is just a regular resistor. The name merely comes from the job it performs. A value of 10 kΩ for the pull-down resistor is common for 5 volt systems.

Similarly, a *pull-up resistor* is connected to the 5 volt connection, as opposed to the ground connection, and keeps the input pin HIGH until brought LOW by the switch. In fact, the Arduino has built-in pull-up resistors on all its I/O pins. They are activated through code.

Using Leaf Switches as Bumpers

Two standard leaf switches mounted to the front of your Teachbot let it detect when it's hit something. With the switches situated to the sides, your bot can determine if the object is on the left or on the right, and so steer around it.

Figure 5-3 shows a pair of leaf switches mounted like bumpers to the front of the robot. Switches like these are available at many online electronics outlets, and are common as surplus.

I haven't augmented the switches with a larger contact area, as for the time being I'm more interested in demonstrating the concepts involved. Use your creativity in enhancing the switches to provide the level of sense detection you want.

Figure 5-3 A pair of leaf switches on the Teachbot. You can attach things to the leaf of each switch to enlarge its contact area.

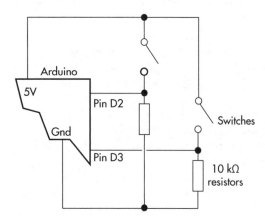

Figure 5-4 Schematic view of connecting two bumper switches to the Arduino microcontroller

To mount each switch, find two suitable holes in the base of your robot, or drill new ones. Most leaf switches have three connections: common, normally open (NO), and normally closed (NC). Wire the *common* and *NO* connections. If space is tight, break off the NC connection to make room.

See Figure 5-4 for a schematic diagram of connecting the two switches to digital pins D2 and D3 of the Arduino. Figure 5-5 shows the same circuit, but in breadboard view.

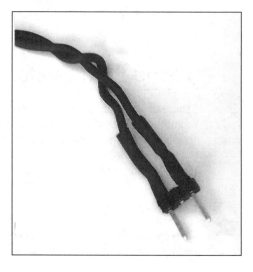

On my prototype I made connectors for the switches by soldering the two wires to pins of a breakaway male header. With these you break off the number of pins you want to use. I cut one connector to three pins wide, removing the middle pin; I soldered the wires from the switch to the outer two pins. For the other connector I cut it to four pins wide, removing the middle two pins.

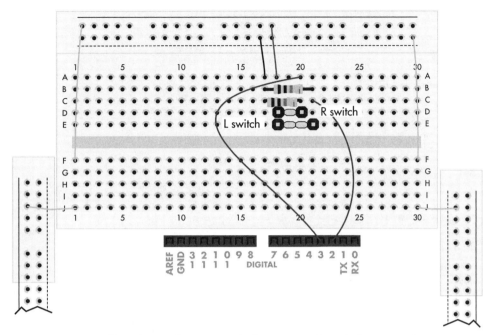

Figure 5-5 Breadboard view of connecting the bumper switches. The layout is only a suggestion. You're free to alter it as you wish.

Important! Make sure all the wires and other components are firmly seated into their breadboard tie-point sockets. Loose connections are the second most common cause for problems when using a solderless breadboard—the most common is plugging the wires into the wrong tie-points!

Refer to the *TeachbotServo_bumper* sketch for a demonstration sketch that shows how to read the switches. The Teachbot sets off going forward until one of its front bumper switches makes contact with an object. The moment the switch closes, the robot quickly reverses direction, then turns in the opposite direction of the obstacle. Time delays are specified in milliseconds: The robot backs up for 500 milliseconds (half a second). It then turns—actually spins—to the right or left for 1500 milliseconds (1.5 seconds).

TeachbotServo_bumper

```
/*
 Teachbot bumper switch demo
*/

#include <Servo.h>

const int ledPin =  13; // Built-in LED
const int bumpRight = 2; // Right bumper pin 2
```

```
const int bumpLeft = 3;   // Left bumper pin 3
int pbLeft = 0;           // Var for left bump
int pbRight = 0;          // Var for left bump
Servo servoLeft;          // Define left servo
Servo servoRight;         // Define right servo

void setup() {
  servoLeft.attach(9);    // Left servo pin D9
  servoRight.attach(10);  // Right servo pin D10

  // Set pin modes
  pinMode(bumpLeft, INPUT);
  pinMode(bumpRight, INPUT);
  pinMode(ledPin, OUTPUT);
}

void loop() {
  forward();              // Start forward
  // Test bumper switches
  pbLeft = digitalRead(bumpLeft);
  pbRight = digitalRead(bumpRight);
  showLED();              // Show LED indicator

  // If left bumper hit
  if (pbLeft == HIGH) {
    reverse();
    delay(500);
    turnRight();
    delay(1500);
  }

  // If right bumper hit
  if (pbRight == HIGH) {
    reverse();
    delay(500);
    turnLeft();
    delay(1500);
  }
}

// Motion routines
void forward() {
  servoLeft.write(0);
  servoRight.write(180);
}

void reverse() {
  servoLeft.write(180);
  servoRight.write(0);
}

void turnRight() {
  servoLeft.write(180);
  servoRight.write(180);
}

void turnLeft() {
  servoLeft.write(0);
  servoRight.write(0);
}
```

```
void stopRobot() {
  servoLeft.write(90);
  servoRight.write(90);
}

void showLED() {
  // Show LED if a bumper is hit
  if (pbRight == HIGH || pbLeft == HIGH) {
    digitalWrite(ledPin, HIGH); // LED on
  } else {
    digitalWrite(ledPin, LOW);  // LED off
  }
}
```

You can experiment with other delay settings, depending on how fast your robot travels. With faster servo motors you can use a shorter delay. The idea is to spin the robot about one-quarter to one-half turn, so it moves away from the obstacle.

 The concept of "left" and "right" (and "front" and "back") are somewhat subjective in a robot like the Teachbot. Either end can be the front, so left and right are relative. If your robot seems to behave opposite to what it should, swap the values in the motion routines (forward, reverse, etc.).

UNDERSTANDING THE BUMPER SKETCH

As with all Arduino sketches, *TeachbotServo_bumper* has three principal parts: declaration, *setup* function, and *loop* function:

- The *declaration* area at the top of the sketch sets up the variables used throughout the program. It also prepares two objects of the *Servo* class. As you read in Chapter 4, "Programming the Teachbot: Making It Move," the *Servo* class is provided as a library that comes with the Arduino programming tools. You use it to control one or more R/C servos. The declaration also defines the two leaf switches as connected to digital pins D2 and D3, and that we'll be using the Arduino's built-in LED (internally connected to pin D13) as a visual indicator.
- In the *setup* function the servos are defined as connected to digital pins D9 and D10. The pins used for the two switches are set as inputs, and the LED is set as an output.
- The main body of the sketch is the *loop* function, which repeats indefinitely. It begins by activating the two servos to move the robot forward. The sketch then uses the *digitalRead* statement to store the current state of the two switches. The instantaneous value of the switches are kept in a pair of variables (*pbLeft* and *pbRight*—the *pb* for pushbutton). These variables are used elsewhere.

Of principal interest in the *loop* function are the two *if* statements. Here's the one that tests the left leaf switch:

```
if (pbLeft == HIGH) {
  reverse();
  delay(500);
  turnRight();
  delay(1500);
}
```

The statement *pbLeft == HIGH* checks to see if the *pbLeft* variable (set earlier based on the state of the left leaf switch) is HIGH. If it is, then the left switch is closed, and the robot has made contact with something. If it's LOW, then the switch is open, and the robot continues on its way.

The *TeachbotServo_bumper* sketch also includes a number of user-defined functions. Most—like *forward* and *reverse*—relate to driving the servo motors. Another function, *showLED,* toggles the LED on pin D13 of the Arduino on or off, depending on whether a switch is closed. Use this as a visual indicator that the programming code is working as it should.

SWITCH TRIGGERS USING POLLING OR INTERRUPTS

The programming in *TeachbotServo_bumper* relies on what's known as *polling:* the sketch repeatedly checks the status of the two switches. If a switch is closed, its value goes from LOW to HIGH; when HIGH the robot is commanded to steer to a new heading. The switches are checked—polled—many times each second.

Polling is an acceptable method when the sketch is relatively simple, and the demands on the Arduino are light. But for code that is more processing intensive, there is a chance the controller will miss when a leaf switch has closed. It'll be busy doing something else, in between polls, and unaware anything has happened.

In truth, you can have a fairly involved sketch and it will still detect 99 percent of all switch closures. The reason: The switch will likely be closed for what are very long periods of time to a microcontroller. For a controller running at 16 MHz, even a brief 100-millisecond (one-tenth-second) contact is like a lifetime, and so in all likelihood the switch closure will be registered.

Still, if you absolutely must ensure that even the most fleeting contact is registered, consider using *hardware interrupts* rather than polling. With an interrupt, special code is run if, and only when, a specific external event occurs. Because the main program loop doesn't have to continually check the state of the pins, it frees up the controller to do other things. Reaction time of an interrupt is measured in microsecond timing, even if the Arduino is busy doing something else. (Actually, this is not always true, depending on how other hardware on the controller is being used. But any additional delay is usually minimal.)

The Arduino Uno supports two hardware interrupts (the Arduino Mega 2560 supports six), internally connected within the Arduino to digital pins D2 and D3. These are the pins that the leaf switches are already connected to, so the only change needed is the software.

See *TeachbotServo_interrupt* for an interrupt-driven example of reading the two bumper switches. Here, the *TeachbotServo_bumper* sketch has been revised to "listen" to a state change on both of the hardware interrupts, with the statements

```
attachInterrupt(0, hitLeft, RISING);
attachInterrupt(1, hitRight, RISING);
```

Note that the interrupts are referred to as *0* and *1*. These correspond to pins D2 and D3, respectively. The labels *hitLeft* and *hitRight* are the functions that are called when the interrupt is triggered. And finally, *RISING* is a built-in constant that tells the Arduino to trigger the interrupt on a LOW-to-HIGH signal transition. This type of transition occurs when the switch closes.

TeachbotServo_interrupt

```
/*
 Teachbot interrupt bumper demo
*/

#include <Servo.h>

const int ledPin =  13;
const int bumpRight = 2;
const int bumpLeft = 3;
volatile int pbLeft = 0;
volatile int pbRight = 0;
Servo servoLeft;
Servo servoRight;

void setup() {
  servoLeft.attach(9);
  servoRight.attach(10);

  // Set pin modes
  pinMode(bumpLeft, INPUT);
  pinMode(bumpRight, INPUT);
  pinMode(ledPin, OUTPUT);

  // Set up interrupts
  attachInterrupt(0, hitRight, RISING);
  attachInterrupt(1, hitLeft, RISING);
}

void loop() {
  forward();     // Start forward
  showLED();     // Show LED indicator

  // If left bumper hit
  if (pbLeft == HIGH) {
    reverse();
    delay(500);
    turnRight();
    delay(1500);
    pbLeft = LOW;
  }

  // If right bumper hit
  if (pbRight == HIGH) {
    reverse();
    delay(500);
    turnLeft();
    delay(1500);
    pbRight = LOW;
  }
}

// Motion routines
void forward() {
  servoLeft.write(0);
  servoRight.write(180);
}
```

```
void reverse() {
  servoLeft.write(180);
  servoRight.write(0);
}

void turnRight() {
  servoLeft.write(180);
  servoRight.write(180);
}

void turnLeft() {
  servoLeft.write(0);
  servoRight.write(0);
}

void stopRobot() {
  servoLeft.write(90);
  servoRight.write(90);
}

void showLED() {
  // Show LED if a bumper is hit
  if (pbRight == HIGH || pbLeft == HIGH) {
    digitalWrite(ledPin, HIGH);
  } else {
    digitalWrite(ledPin, LOW);
  }
}

// Interrupt handlers
void hitLeft() {
  pbLeft = HIGH;
}
void hitRight() {
  pbRight = HIGH;
}
```

Both *hitLeft* and *hitRight* set their corresponding "pb" (pushbutton) variables to HIGH. The program then immediately exits the interrupt handler. The next time the Arduino repeats its *loop* it notices that a pushbutton is HIGH, and performs the needed obstacle avoidance maneuver. (Note also that the pushbutton value is manually set back to LOW, in anticipation of the next bump.)

Check out the use of the *volatile* qualifier for the *pbLeft* and *pbRight* variables. *Volatile* tells the Arduino compiler to use random access memory (RAM) to store these values, rather than one of the 32 internal working registers built into the ATmega chip. Under some conditions, data placed in a working register, and written to or accessed by an interrupt, may not be accurate.

What's a working register? It's like a variable, but it's built into the hardware of the microcontroller. It's used, when possible, instead of RAM to conserve memory space.

You might be asking why the code to control the servos isn't in the interrupt handlers. The reason is this: The *delay* statement, which is used to steer the robot around an obstacle, is disabled while in an interrupt. And in any case it's usually best not to place time-intensive functionality within interrupt handlers.

TO LET BOUNCE, OR DEBOUNCE?

In a perfect world mechanical switches would produce clean, reliable digital signals for our microcontrollers. Alas, nearly all mechanical switches suffer something called *switch bounce*—instead of a nice LOW-to-HIGH digital pulse when a switch closes, there might be five, ten, maybe even dozens of irregular glitches, all caused as the metal contacts in the switch settle into position. This happens very quickly, usually in just a few milliseconds.

For some applications it's absolutely necessary to *debounce* the output of the switch. In debouncing, all the glitches are removed, providing the microcontroller with that single sweet pulse we want. One way to debounce is with an added 0.01 or 0.1 µF capacitor across the terminals of the switch. The capacitor acts to delay the rise and fall of the switch signal, effectively removing the glitches.

You can also do it in software, typically using delays so that the microcontroller ignores all but the first signal transition. Both the bumper and interrupt examples noted earlier don't directly use switch debounce. Software delays are already built into the code, plus R/C servos are very slow creatures and don't react fast enough for bounce to be a problem.

Should you ever need to debounce your switch inputs there's a separate class library you can download and use with the Arduino. The library is called *Bounce,* and it's available from the main Arduino Language reference pages. There's also a *Debounce* code example that comes with the Arduino programming IDE.

MOUNTING ALTERNATIVES, MORE SWITCHES

A few quick notes before moving on. So far I've talked about the two switches in the front of the robot, situated right and left. Feel free to put switches anywhere you want. You might instead have a front and back switch, or a bunch of switches all around the periphery of the robot.

While the concept is not illustrated for the Teachbot, you can use more than two switches. When using the Arduino Uno (and similar boards), you'll have to rely on polling, as there are only two pins that support hardware interrupts.

Let There Be Light (And Let Your Teachbot See It!)

Next to tactile feedback, reacting to light is the most common robotic sense. In lieu of actual vision, the robot uses simple electronic components such as photoresistors and phototransistors that are sensitive to light. Your bot may react to the absence or presence of light, or it may measure the brightness, color, or other qualitative aspect of the light.

Photoresistors—also called photocells, light dependent resistors, or CdS (for cadmium sulfide) cells—are perhaps the easiest to use as basic light sensors. The photocell is a resistor whose value changes depending on the amount of light that strikes its sensing surface. In darkness, the photocell has a high resistance, typically in the neighborhood of 100 kΩ to over

1 megohm, depending on the component. The resistance falls as more light strikes the cell. In high brightness the resistance may be as low as 1 kΩ to 10 kΩ.

The exact dark/light resistance values differ depending on the component, and even among photocells of the same make and model. Typical value tolerance is 10 to 50 percent. You can purchase photocells new, but they're common finds in the surplus market.

Get a variety pack, and use your multimeter to "grade" each one. Cracks and other injury spell doom to a photocell; air and moisture degrade the sensing surface, rendering it useless. Toss any that don't react properly to a nearby desk lamp.

CONVERTING PHOTOCELL OUTPUT TO A VOLTAGE

Because a CdS photocell behaves like a resistor, you can convert its output to a varying voltage merely by connecting another resistor to it in series, as shown in Figure 5-6. The value of the series resistor depends on the dark/light resistance range of the photocell, and how you want to use it. The cells I used had a dark resistance of about 40 kΩ and a light resistance of 30 ohms. In average room brightness, the cells had a 10 kΩ resistor, so I selected a 10 kΩ series resistor.

The voltage at the point between the photocell and series resistor is a ratio of the two resistance values. With the two resistances equal, the divided voltage between them is one half of the supply voltage; in the case of 5 volts, in average room light the output voltage is 2.5 volts. The voltage decreases in darkness, increases as more light strikes the photocell.

You will need to experiment with the series resistor to determine its best value, based on the specific photocells you use. Or if you want to get fancy, you might want to try a 50 kΩ or 100 kΩ potentiometer in place of a fixed resistor, allowing you to fine-tune the series resistance as needed.

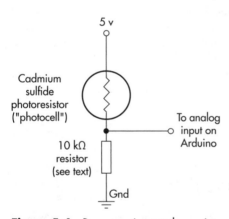

Figure 5-6 By connecting another resistor in series with a cadmium sulfide photoresistor, the output is converted to a varying voltage. In this particular arrangement, the voltage increases under stronger light.

DEMONSTRATION OF BASIC LIGHT SENSITIVITY

Refer to *TeachbotServo_SimpleCds,* a basic sketch that tests the operation of the photocell. Wire the photocell as shown in Figure 5-6, and connect the output to analog pin A0 of your Arduino. Compile and upload the sketch, then open the Serial Monitor window. You'll see a series of numbers; they correspond to the output voltage of the sensor converted to a 10-bit (0 to 1023) numeric value. You should get a low number when all light to the photocell is blocked, and a higher number under full illumination.

TeachbotServo_SimpleCds

```
/*
 Teachbot CdS cell demo
*/

int cds = 0;

void setup() {
  Serial.begin(9600);
}

void loop() {
  // Read analog pin A0 and display value
  //  on Serial Monitor window
  cds = analogRead(A0);
  Serial.println(cds, DEC);
  delay (200);
}
```

STEERING YOUR ROBOT WITH A FLASHLIGHT

By using two photocells mounted on each side of your Teachbot you can literally steer it by flashlight. Under just room light the robot is set to stop, waiting for your command. Aim the flashlight so that light falls more or less equally on both photocells, and the robot will move forward. When the light levels aren't equal, the robot will turn toward the photocell that has more light falling on it.

Refer to Figure 5-7 for a schematic diagram of the two-photocell setup. Figure 5-8 shows the same circuit but in breadboard view. For the prototype Teachbot I attached the photocells directly to the solderless breadboards, stiffening the leads with unshrunk heat-shrinkable tubing. This arrangement allows for easy positioning of the eyes—upward and slightly outward are good orientations to start. Figure 5-9 shows the photocell "eyes" atop the Teachbot.

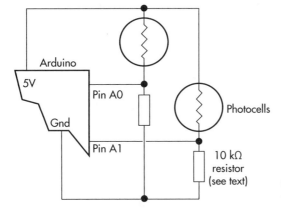

Figure 5-7 Schematic view of connecting two photocells to the Arduino microcontroller

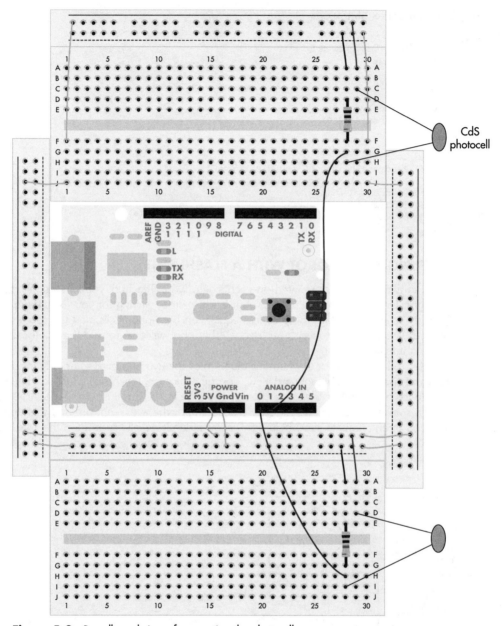

Figure 5-8   Breadboard view of connecting the photocells

Bear in mind light can strike the photosensitive surface of the cell from the rear. You may want to add a layer or two of black tape to prevent light spoilage.

Figure 5-9 Mount the photocells on the top of the Teachbot, directly onto its two main solderless breadboards. This allows the bot to follow a flashlight beam.

Check out *TeachbotServo_lightsteer,* a sketch that uses the current values of the photocells to make quick steering adjustments to the left or to the right. In the declarations area the code

```
const int ambient = 600;
const int threshold = 800;
```

sets two comparison values used elsewhere in the sketch. *You will need to experiment with these values depending on the room environment and photocell characteristics!* These values worked for me; you can start with them, but expect to try other values as you fine-tune the performance of the steering.

ambient sets the upper level of just the ambient (natural) light in the room. This is the amount of light that hits the photocells under normal lighting conditions. For me, the ambient light value was about 520 to 530, so I made it a little higher (600) for extra headroom.

threshold sets the lower level of the light beamed from your flashlight. I set the value at 800 based on using a nine-LED flashlight (with fresh batteries) 1 to 2 feet away from the robot. For best results, use a bright flashlight near the robot—the farther away you get, the less light that falls on the photocells.

TeachbotServo_lightsteer

```
/*
 Teachbot steering by light demo
*/

#include <Servo.h>

// CdS cell reference values
//  (you need to experiment)
const int ambient = 600;
const int threshold = 800;
```

```
int lightLeft = 0;
int lightRight = 0;

Servo servoLeft;
Servo servoRight;

void setup() {
  servoLeft.attach(9);
  servoRight.attach(10);
}

void loop() {
  // Read light sensors connected to
  //   analog pins A0 and A1
  lightLeft = analogRead(A1);
  lightRight = analogRead(A0);

  // Stop robot if below ambient
  if (lightRight < ambient || lightLeft < ambient) {
    stopRobot();
  } else {
    forward();
    // Steer to Left if right CdS below threshold
    if (lightRight < threshold) {
      turnLeft();
      delay (250);
    }
    // Steer to right if left CdS below threshold
    forward();
    if (lightLeft < threshold) {
      turnRight();
      delay (250);
    }
  }
}

// Motion routines
void forward() {
  servoLeft.write(0);
  servoRight.write(180);
}

void reverse() {
  servoLeft.write(0);
  servoRight.write(180);
}

void turnRight() {
  servoLeft.write(180);
  servoRight.write(180);
}

void turnLeft() {
  servoLeft.write(0);
  servoRight.write(0);
}

void stopRobot() {
  servoLeft.write(90);
  servoRight.write(90);
}
```

Following a Line

You want simple. You want cheap. You want effective. And, of course, you want fun.

That pretty much sums up line following, probably the simplest of all navigation systems for mobile robots. Even high-tech factories use line following, where a predefined path is marked on the ground. The path can be a painted black or white line, or it might be a wire buried beneath a carpet.

You can easily and inexpensively incorporate a tape-track navigation system in your robot. The line-following feature can be your bot's only means of intelligence, or it can be just one part of a more sophisticated machine. You could, for example, use the tape to help guide your robot back to its battery charger nest. Beats clicking your heels three times and saying, "There's no place like home. . ."

THE BASICS OF LINE FOLLOWING

With a line-following robot you place white, black, or reflective tape on the floor—the color of the tape is selected to contrast with the floor. Using simple optics, the robot then follows that line.

Black PVC electrical tape absorbs infrared light, and it's cheap and easy to use. So the most popular method of making a track is to lay down a path of electrical tape onto white paper, like that in Figure 5-10. I prefer using a stiff poster board that's coated with a smooth finish on at least one side. The coating reflects more light, and helps prevent the paper from being transparent to the infrared light. Try the nearby dollar store for inexpensive poster board. I get mine for 50 cents a sheet. Each sheet measures 22 by 28 inches. It comes in colors, but you want to start out with white.

For the best results, the floor should be hard, like wood, concrete, or linoleum, and not carpeted. You can also use a large table. To sense the line, optical sensors are placed in front of, or under, the robot (see Figure 5-11). These sensors incorporate an infrared LED (*emitter*) and an infrared phototransistor (*detector*). The output of the phototransistor indicates whether or not it sees light reflected off the floor. Assuming a black line on a white floor, the absence of reflected light means the robot is over the line.

You can purchase ready-built electronics for use in line following, or make your own. The cost is comparable, though the home solution offers a little more flexibility. I'll begin with a discussion of using ready-built emitter/detector optics, then cover a homebrew method you can try.

Regardless of the electronics you use, reflective sensors used for line following like to be as close to the ground

Figure 5-10 Black PVC electrical tape on white poster board makes an excellent line-following course.

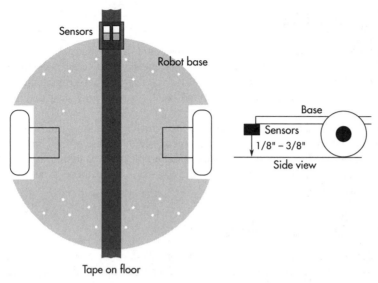

Sensors

Robot base

Base

Sensors

1/8" – 3/8"

Side view

Tape on floor

Figure 5-11 Placement of the reflective sensors on the robot. For best results the face of the sensors should be 1/8" to 3/8" from the ground.

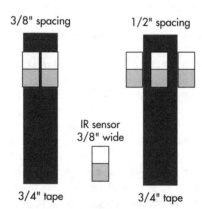

3/8" spacing

1/2" spacing

IR sensor
3/8" wide

3/4" tape

3/4" tape

Figure 5-12 The spacing between reflective sensors is dependent on the width of the line. The line must be visible to at least one sensor at any time.

as possible, while not making actual contact. The ideal distance from the sensor to the ground is 1/8" to 3/8", depending on the design of the sensor. I mounted the two sensors to the underside of the Teachbot using plastic spacers, in order to decrease the distance from sensor surface and floor.

Whether you use homebrew line-following sensors, or rely on ready-made modules, the emitter and detector can be aligned either in column or row orientation, whatever fits best. By their nature, line following requires at least two pairs of emitters/ detectors. These are placed a specific distance from one another, in order to determine the location of the line under the robot.

The spacing of the emitter/detector pairs depends on the width of the tracking tape. To be effective, at least one sensor should detect the line at any one time. With too much space, a thin line can get "lost" between sensors. You can fix this either by grouping the emitter/detector pairs closer together or using a wider tape. Figure 5-12 shows some variations. As most PVC electrical tape is 3/4" wide, a spacing of 3/8" to 1/2" is most common.

When I talk of spacing, I mean the distance between the *optical centers* of each emitter/ detector pair, regardless of the physical dimensions of the sensor. Given sensors that are physically 3/8" wide, and 3/8" between optical centers, the modules are actually right next to each other.

USING READY-BUILT LINE-FOLLOWING ELECTRONICS

Several online sources offer what I refer to as *unitized emitter/detector modules*. These include modules from SparkFun, Pololu, Parallax, and others. All are provided on a small circuit board, where all the necessary components are already mounted. Simply attach wires to the module: power, ground, and signal. A single hole allows you to mount the module most anywhere on the robot.

A variation of this concept is the *emitter/detector array*, which combines two or more pairs of infrared LEDs and phototransistors on a single circuit board. All components are already on the board, and the LED/phototransistor pairs are spaced 3/8" to 1/2" from one another, ideal for line following.

One such array is described in this book. For details see Chapter 9, "Enhancing the Teachbot."

For this example I'm using a pair of SparkFun QRE1113 analog line sensor breakout surface mount boards. The QRE1113 is a reference to the infrared emitter/detector device at the heart of the board. The signal output is analog—it's a varying voltage depending on the amount of light striking the detector. In the case of the SparkFun board, the less light reflected, the lower the voltage.

Refer to Figure 5-13 for a connection diagram, and Figure 5-14 for how the two sensors are mounted side by side on one end of the Teachbot. The spacing between the sensors is

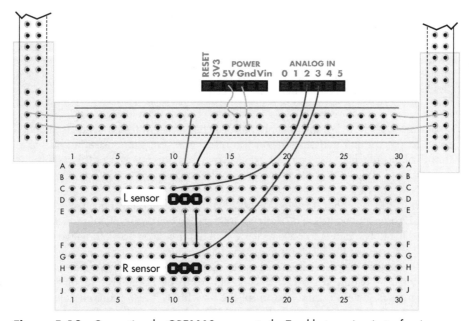

Figure 5-13 Connecting the QRE1113 sensors to the Teachbot requires just a few jumper wires. Use a three-wire servo extension to mate the sensor modules with a three-pin long header seated in the breadboard.

Figure 5-14 A pair of SparkFun QRE1113 sensor modules, attached to the front underside of the Teachbot using 4-40 hardware and small spacers. The spacers should place the optics of the module to within about 1/8" of the ground.

ideal when using 3/4" wide black electrical tape. Notice that the QRE1113 breakout boards have three-pin headers, and are connected to the Teachbot's breadboard and Arduino via a standard three-wire servo extension. Opt for an 8" to 10" extension.

SparkFun also offers a *digital* version of the QRE1113 breakout board. Rather than a varying voltage, its signal is a simple off/on, depending on the brightness of the surface. While these are perfectly acceptable sensors, they won't work for this example. You need the analog kind.

If you use an alternative emitter/detector module be sure it provides an analog output. Check the specifications for the module.

Because the "front" of the Teachbot already has a pair of feeler switches attached to it, I've opted to put the infrared sensors on the other end. For demonstrating line following, the "front" of the Teachbot will be reversed. As you'll see, this involves nothing more than altering a few lines of code so that the motors go in the opposite direction for forward, reverse, and turns.

Refer to the *Teachbot_LineFollowQRECalibrate* sketch for a quick example of using the QRE1113 breakout boards. The sketch is intended to help you calibrate the threshold between dark and light. The aim is to have as large of a difference as possible between readings when the sensors are over the black tape and the white paper.

You'll find that the closer the sensor is to the floor, the better. In my tests the sensor had to be mounted to within just a few millimeters from the floor. This required adding spacers between the modules and the bottom of the Teachbot. It also required adjusting the skid on the opposite end to prevent the robot from tipping side to side. Too much tipping increases the distance between sensor and ground, and the difference in the dark/light readings isn't enough for reliable operation.

Teachbot_LineFollowQRECalibrate

```
const int irLeft = A2;
const int irRight = A3;

void setup() {
  Serial.begin(9600);
}

void loop() {
  Serial.print(analogRead(irLeft), DEC);
  Serial.print("\t");
  Serial.println(analogRead(irRight), DEC);
  delay(100);
}
```

Use the *Teachbot_LineFollowQRECalibrate* sketch to find an average threshold value between the dark and light readings. For example:

Over Black Tape	Over White Background
70	870

Select a threshold value somewhere between these two, favoring the higher end of the scale, such as 650. Use this threshold in *Teachbot_LineFollowQREDemo,* which demonstrates basic line-following functionality. Make a line-following course using one or two large pieces of construction paper. Use a new, fresh roll of 3/4" black electrical tape to produce an oval-shaped track—keep the course simple for now; you can get fancy later.

Don't stretch the tape as you lay it down, or else it'll shrink over time, and cause the line to "creep" and change position. Use your finger to smooth out any puckered-up edges on the corners of the track. The track doesn't have to be perfectly flat with the paper, but there shouldn't be any obvious buckling or rough spots to snag on the Teachbot's parts.

Teachbot_LineFollowQREDemo

```
#include <Servo.h>

Servo servoLeft;            // Define left servo
Servo servoRight;           // Define right servo
const int lineLSense = A3;
const int lineRSense = A4;
int irReflectR = 0;
int irReflectL = 0;
int thresh = 650;

void setup() {
  servoLeft.attach(9);    // Left servo pin D9
  servoRight.attach(10);  // Right server pin D10
}
```

```
void loop() {
  // Read reflective sensors
  irReflectL = analogRead(lineLSense);
  irReflectR = analogRead(lineRSense);

  if (irReflectL >= thresh && irReflectR >= thresh) {
    line_forward();   // on line
  }

  if (irReflectL >= thresh && irReflectR <= thresh) {
    line_spinLeft();  // veering off right
    delay(4);
  }

  if (irReflectL <= thresh && irReflectR >= thresh) {
    line_spinRight();  // veering off left
    delay(4);
  }

  // If line is lost try to reacquire
  if (irReflectL < thresh && irReflectR < thresh) {
    line_spinRight();
    delay(20);
  }
}

// Motion routines for line following
void line_forward() {
  servoLeft.write(0);
  servoRight.write(180);
}

void line_slipRight() {
  servoLeft.write(90);
  servoRight.write(180);
}

void line_slipLeft() {
  servoLeft.write(0);
  servoRight.write(90);
}

void line_spinRight() {
  servoLeft.write(180);
  servoRight.write(180);
}

void line_spinLeft() {
  servoLeft.write(0);
  servoRight.write(0);
}
```

The logic of the sketch is fairly simple:

- If the right sensor doesn't detect the line, it means the robot has veered too far to the right. To compensate the robot steers back over to the left.
- Conversely, if the left sensor doesn't detect the line, the robot corrects its path by briefly steering back toward the right.

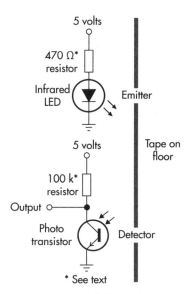

Figure 5-15 The basic line-following circuitry has two parts: emitter and detector. Light from the emitter reflects off a surface and into the detector.

Notice that the servo directions are reversed for all the motion routines you've previously practiced with. This is because the robot is set to run in the opposite direction. This is perfectly acceptable for a two-wheeled robot like the Teachbot, where there is no "official" front and back.

MAKING YOUR OWN LINE-FOLLOWING ELECTRONICS

The ready-built line-following modules are easy to use, but aren't as flexible. You can't easily modify the components used on the module, so you can't change things like the sensitivity or range. These kinds of adjustments are possible if you make your own line-following modules.

You need an infrared LED and a phototransistor to build a line-following sensor. Mount these two components close to one another, and point them in the same direction. All you need to finish the circuit is a couple of resistors, as shown in Figure 5-15. You can mount the resistors with the LED and phototransistor, or locate them on the breadboard.

Let's take a closer look at how these components are used to make a line-following circuit.

Selecting the Emitter LED and Current-Limiting Resistor
The electronics for line following are technically two separate circuits, connected to the same power source. The first part is the emitter LED and its resistor. The purpose of the resistor is to limit current flowing through the LED. You select the value of the resistor based on how much current you want to pass through the LED. The more current, the brighter the LED will glow.

Ideally, the LEDs you use come with specifications, so you know technical details like *forward voltage drop* and *maximum forward current*. With this info, you can use some simple math to calculate the value of the resistor you need. Here's the formula:

$R = V_{in} - V_{drop} / I_f$

V_{in} is 5 volts. Assuming V_{drop} is 1.7 volts (kinda typical for infrared LEDs), and a desired 30 mA (milliamps) current, the formula becomes:

$110 = 5 - 1.7 / 0.030$

If the result isn't a standard resistor value, always choose the next highest.

(Notice I_f is given as a decimal fraction. The formula expects amps, but these LEDs are rated in milliamps. You need to move the decimal point over to convert amps to milliamps: a value of 0.030 is 30 milliamps.)

If you like to buy from surplus, like I do, you may not have any specs for the infrared LEDs you use. Instead, you can apply some educated guesswork to select a resistor value that's close to what you need. I usually start with a 470 ohm resistor, as this provides enough drive current to allow the LED to glow, but not so much that it will burn out the LED.

Depending on the LED, you may need to reduce the value of the resistor, knowing that if you get too low (say, under 150 ohms or so), the chances of burning out the component dramatically increase.

Many IR LEDs are engineered for fairly high maximum forward current, so you have a bit of wiggle room. While the average visible-light LED might have a maximum forward current of 30 mA, the typical IR LED is often rated higher, some 50 mA or more. (Of course, there are always exceptions. For the sake of science you should be prepared to blow out an LED now and then!)

If you have a choice, pick an LED with a narrow viewing angle. It shines more light into a spot. The generic T-1 size (3mm diameter) water-clear LED is a good choice. These have a built-in lens, and are both plentiful and cheap. They're also easy to mount by drilling a 1/8" diameter hole into the bottom of the robot.

Selecting the Phototransistor and Divider Resistor

The second half of the line-following circuit is composed of a phototransistor, along with a resistor that forms a voltage divider. This is what produces a varying voltage at the output of the sensor. The value of the resistor depends on the characteristics of the phototransistor and the circuit you're connecting to; the higher the value of the resistor, the more sensitive the transistor becomes to variations in light intensity.

The easiest way to determine the value of the resistor is through empirical testing: Wire the phototransistor as shown in Figure 5-16, starting with a 10 kΩ resistor. Connect a voltmeter between the Output and ground connections, as shown. Use an incandescent (not fluorescent or LED) desk lamp to alternately shine or block light to the phototransistor. You should see the voltage fluctuate on your meter. It may not be much.

Now try incrementally larger resistance values until you observe a good *voltage swing*—the difference in voltage between darkest and lightest conditions. It won't be the full 0 to 5 volts of the power supply, but the larger the difference, the better. You don't need (or want) to select ever-higher resistance values when a lower one works just as well. For the phototransistors

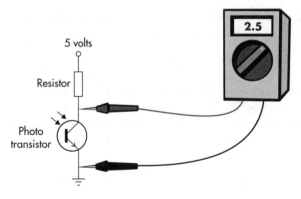

Figure 5-16   Find the optimum value for the resistor used with the phototransistor by testing with a digital voltmeter. Select a resistor that provides a good voltage swing between dark and light.

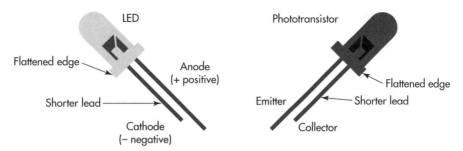

Figure 5-17 Detail of the typical LED and phototransistor T-style packages. Use the short lead and/or flattened side to determine the correct pinout.

I used for my prototype Teachbot, I found good sensitivity and voltage swing with 100 kΩ resistors.

But what if the phototransistor still doesn't respond well to light, even when using very high values? The phototransistor may be bad, or it may be connected in reverse. Try flipping it around in the circuit, or select another one. See Figure 5-17 for how to determine the leads of the typical LED and phototransistor. Note that in a phototransistor the flattened side/short lead is usually the collector.

Like LEDs, phototransistors come in a variety of package styles. Look for the same package type you use for the LED. A phototransistor in a T-1 (3mm diameter) package has its own built-in lens, and it's easy to mount through a hole in the bottom of your robot.

ARRANGING THE INFRARED LED AND PHOTOTRANSISTOR

The LED emitter and phototransistor detector should be mounted side by side and pointing in the same direction. To avoid light "crosstalk"—light that shines from the side of the LED and directly into the phototransistor—add some black heat-shrink tubing around the LED, the phototransistor, or both. Or place a small piece of metal or opaque plastic between the two to act as a baffle.

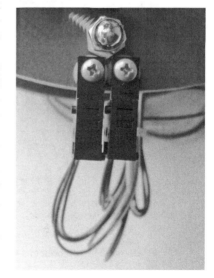

While using separate LEDs and phototransistors is the least expensive method of creating a line-following sensor, an easier approach is to use an all-in-one reflective sensor. These come in various shapes and sizes, and are fairly common on the surplus market.

Shown here is a pair of reflective sensors mounted on the outside edge of the Teachbot. These sensors, typical of surplus finds, provide their own mounting flange. They're easy to use, but because this style of reflector is bulky, they can't be mounted under a low-profile robot such as the Teachbot. Instead, they have to be placed on the outside edge. That's okay, as it demonstrates one of many ways to perform line following.

CONSTRUCTING A TWO-SENSOR LINE FOLLOWER

Figure 5-18 shows the schematic view of connecting two IR LEDs and two phototransistors to the Arduino. Since the output of the phototransistors is a variable voltage, we use two of the Arduino's analog-to-digital converter inputs, specifically pins A2 and A3. Figure 5-19 shows the same circuit, but in breadboard view.

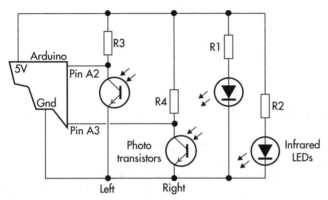

Figure 5-18 Circuit for connecting two LEDs and phototransistors to the Arduino. See the text for the values of the four resistors.

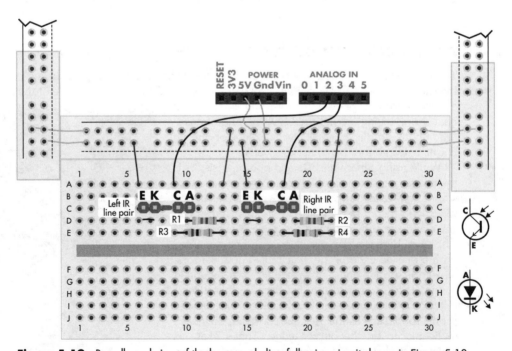

Figure 5-19 Breadboard view of the homemade line-following circuit shown in Figure 5-18

Resistor values to start (see the previous discussion for selecting the values based on the LEDs and phototransistors you use):

R1, R2: 470 ohms
R3, R4: 100 kΩ

You can use the same *Teachbot_LineFollowQREDemo* line-following sketch as presented earlier in this chapter. Experiment with the value of the *thresh* variable to obtain the best results.

- Line is detected by both sensors: keep going forward
- Line is detected only by the left sensor: steer briefly to the left
- Line is detected only by the right sensor: steer briefly to the right
- Line is not detected by either sensor: steer in a circle, and attempt to reacquire the line

If your robot goes the wrong way you can simply swap the servo connections on the breadboard—left becomes right, and right becomes left.

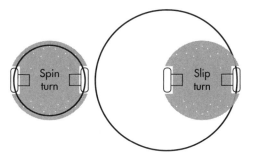

Feel free to experiment with the sketch to see how to improve the performance of your line-following Teachbot. I've intentionally kept it simple to make it easier to understand. Try using the *line_slip* functions rather than the *line_spin* functions. How does this affect maneuverability of the robot?

Chapter 6

Programming the Teachbot: Letting It Explore

You've spent hours designing and building your latest robot creation. You bring it into your living room, fire it up, and step back.

Behold: Your beautiful new robot smashes into the fireplace and scatters itself into tiny pieces over the living room rug. You remembered things like motor speed controls, colorful blinky lights, even a synthetic voice, but you forgot to provide your robot with the ability to look before it leaps.

In previous chapters you learned how to give your Teachbot the sense of touch and light. Augmenting these basic senses are methods to detect objects in *proximity* to your robot, seeing what's there without having to actually bump against them.

Proximity detection forms the basis of *collision avoidance*—how to keep your bot from crashing into things in the first place. Collision avoidance takes many forms. Some collision avoidance techniques are designed to detect objects very close to the robot, while others are made to detect objects several feet away.

In this chapter you'll learn about two popular forms of proximity detection—ultrasound and infrared—and how these low-cost sensors are interfaced and used with the Arduino. Plus, you'll discover how to add a rotating turret so the Teachbot can scan the room to look for things nearby.

 Be sure to visit the ARB Support Site (see Appendix A) for additional code examples and variations. All sketches from this book are available for ready download.

FYI Brand new to Arduino programming? Then be sure to first read through Chapter 2, "Arduino Up and Running." And, of course, refer to the official Arduino Language reference pages at *www.arduino.cc*.

Understanding Non-contact, Near-Object Detection

In Chapter 5 you learned how mechanical switches are used as a form of touch sensor to detect contact with objects. Contact detection provides an immediate signal that something looms directly in the way.

Non-contact detection senses objects without first having to hit them. And *near-object detection* does just what its name implies: It senses objects that are close by, from perhaps just a breath away to as much as eight or ten feet. These are objects that a robot can consider to be in its immediate space; objects it may have to deal with, and soon. These objects may be people, animals, furniture, or other robots.

By detecting them, your robot can take appropriate action, which is defined by the programming you give it. Your bot may be programmed to come up to people and ask for their name. Or it might be programmed to run away whenever it sees movement. In either case, it won't be able to accomplish either behavior unless it can detect the objects in the neighborhood.

There are two common methods of achieving near-object detection: proximity and distance:

- *Proximity* sensors care only that some object is within a zone of relevance. That is, if an object is near enough to be considered important. Objects beyond the proximal range of a sensor are effectively ignored because they are not seen. Out of view, out of mind.
- *Distance measurement* sensors determine the space between the sensor and whatever object is within detection range. Distance measurement techniques vary; almost all have notable minimum and maximum ranges. Few yield accurate data if an object is smack-dab next to the robot, or very far away.

Collectively these sensor types are often referred to as *rangefinders,* though only a device that actually measures and reports the distance of the covered range is a true rangefinder.

Among the most common proximity and distance measurement detectors used in robotics are ultrasonic transducers and specialty infrared sensors made by Sharp. Depending on the design of the specific sensor, either can be used for proximity *or* distance measurement. But in practice, the Sharp IR sensors are best suited for proximity, and ultrasound sensors are the ideal choice for measuring distance. That's how these two detectors are used in the Teachbot.

Using an Ultrasonic Distance Sensor

Ultrasonic distance measurement—also called *ultrasonic ranging*—has been around for decades. It uses high-frequency sound outside the range of human hearing; the typical frequency is around 40 kHz. The sound is emitted as a quick burst from a *transducer,* a kind of speaker. The sound bounces off an object, and the echo is received by the same or another transducer.

A circuit then computes the time it took between the transmit pulse and the echo, and comes up with distance. While the process may sound complex, it's actually very easy to do with a ready-built ultrasonic distance sensor and microcontroller like the Arduino.

At sea level, sound travels at a speed of about 1130 feet per second (about 344 meters per second) or 13,560 inches per second. This time varies depending on atmospheric conditions, including air pressure, temperature, and humidity. The time it takes for the echo to be received is in microseconds if the object is within a few inches to a few feet of the robot. The overall time between transmit pulse and echo is divided by two to compensate for the round-trip travel time between robot and object.

Given a travel time of 13,560 inches per second for sound, it takes 73.7 µs (microseconds or 0.0000737 seconds) for sound to travel 1 inch. If an object is ten inches away from the ultrasonic sensor, it takes 737 µs to travel there, plus an additional 737 µs to travel back, for a total "ping" time of 1474 µs. The calculation is

```
(1474 / 74) / 2 = 10
```

First, divide the total transit time by 74 (a whole number is used to avoid floating-point math), then divide by 2. The result is the distance from sensor to object, in inches.

> You can substitute 74 with 29, which gives you the distance in centimeters. Example:
>
> ```
> (1474 / 29) / 2 = 25
> ```
>
> Assuming you're working with whole numbers only, providing values in centimeters gives you a little more resolution—2.5 times more, to be exact. Just remember to be consistent in your sketches. Avoid mixing inches and centimeters, or else your robot's distance calculations will be significantly flawed.

For the Teachbot I've selected the Parallax Ping. It's relatively low cost and easy to use, requiring just one signal line to the Arduino, in addition to power and ground. There are, of course, other ultrasonic rangers that do pretty much the same thing. You can substitute if you'd like. There's the Devantech SRF05, for example, and various models by Maxbotics. All are excellent products and are well supported by their manufacturers.

Figure 6-1 is a schematic diagram for connecting the Ping to the Arduino, and Figure 6-2 shows the same circuit in breadboard view. Hookup is simple—just route wires between the sensor and microcontroller. If you're using an SRF05 or similar, it comes to you with plated-through holes for the electrical connections. You can solder wires directly to the board, or better yet use a five-pin male header. You can solder the header from either side. I like attaching the wires from the rear (opposite the transducers), so that they don't get in the way of the business end of the sensor.

> **FYI**
>
> This is the final chapter that I'll be providing both breadboard wiring diagrams and schematics. In subsequent chapters I'll show how to hook up the circuits using just a schematic view. I tend to use a minimum of special symbols, and most things are self-explanatory, so even if you're new to schematics you really shouldn't have trouble seeing what goes where.

Be careful to not cross up the power and ground wires. Review the datasheet for the correct pinout. You can use a three-wire female-female servo extension cable to connect the Ping to the breadboard on your Teachbot.

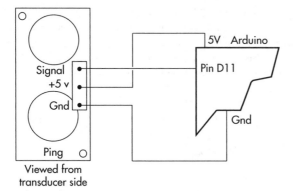

Figure 6-1 Connection schematic for the Parallax Ping ultrasonic ranging module

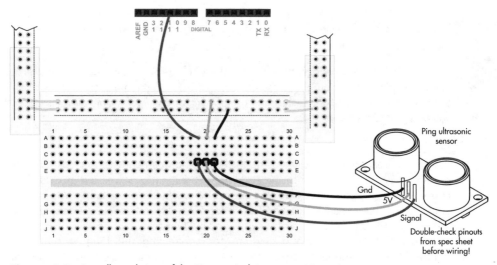

Figure 6-2 Breadboard view of the Ping-to-Arduino connection

DEMONSTRATING BASIC FUNCTIONALITY

Refer to the *Teachbot_Ping* sketch for a demonstration sketch that shows how to take a reading from the Ping sensor, and display it in the Serial Monitor window. The value returned by the *readSonar()* function is converted to whole inches. According to its documentation, the Ping can measure distances from 2 cm to 3 meters. That equates to a range of 1 inch (rounded up) to 118 inches.

Teachbot_Ping

```
const int pingPin = 11;
unsigned int duration, inches, cm;

void setup() {
  Serial.begin(9600);
}
```

```
void loop() {
  Serial.println(sonarRead(), DEC);   // Show in Serial Monitor
  delay(200);
}

int sonarRead () {
  pinMode(pingPin, OUTPUT);            // Set pin to OUTPUT
  digitalWrite(pingPin, LOW);          // Ensure pin is low
  delayMicroseconds(2);
  digitalWrite(pingPin, HIGH);         // Start ranging
  delayMicroseconds(5);                //    with 5 microsecond burst
  digitalWrite(pingPin, LOW);          // End ranging
  pinMode(pingPin, INPUT);             // Set pin to INPUT
  duration = pulseIn(pingPin, HIGH);   // Read echo pulse
  inches = duration / 74 / 2;          // Convert to inches
  cm = duration / 29 / 2;              // Convert to centimeters
  return(cm);
}
```

Place your hand at various distances in front of the Ping to verify the readings displayed in the Serial Monitor window. Check the accuracy of the readings using any flat object, such as a book, and a ruler or tape measure. Try to keep the book perpendicular to the sensor. You'll find accuracy depends on the surface of the object and its distance from the sensor. Hard, smooth objects tend to reflect sound better, so the measurements are more accurate. The actual measured distance may be off by as much as 5 to 15 percent, which is normal for low-cost ultrasonic rangers.

MOUNTING THE SENSOR

To test the Ping, you can just plug it into the solderless breadboard, but for a working bot you'll want more solid mounting. The Ping enjoys a standard size for ultrasonic sensors, and you can buy or make a mounting bracket for it. Figure 6-3 shows the Ping mounted to an aluminum bracket.

Figure 6-3 The Ping ultrasonic ranger module attached to a bracket and mounted on the front of the Teachbot

This bracket is included as part of a rotating turret kit, described later in this chapter. The turret is operated by an R/C servo, which comes with the kit. You can also use a universal L-bracket, which Parallax also offers. It's a less expensive alternative if you don't plan on motorizing the Ping. The turret kit and L-bracket work with many other brands of ultrasonic sensors that share the same physical dimensions as the Ping.

When mounting the Ping—or any sensor for that matter—to a metal bracket, be careful to avoid short circuits. Don't let any part of the circuit board touch the bracket, or damage could result. Use spacers (preferably nylon or other nonconductive material) to physically separate the Ping from its mounting bracket.

USING AN ULTRASONIC SENSOR TO AVOID OBSTACLES

See *Teachbot_PingAvoider*, on the ARB Support Site, for an example sketch of using the Ping ultrasonic ranger to check for obstacles ahead of the Teachbot. If the robot comes to within five inches of an object straight ahead, it will stop, turn to orient itself in a new direction, then move forward again.

The Ping example is combined with the interrupt-driven switch detection sketch described in Chapter 5, "Programming the Teachbot: Seeing It React." It's always a good idea to use multiple sensors to detect obstacles. If, for some reason, the ultrasonic sensor misses an object straight ahead, the robot will still have the two touch switches as a backup.

This is more than just insurance; in actual practice, you may discover your ultrasonic sensor misses objects now and then. Given a completely stationary robot and stationary objects, an ultrasonic sensor is likely to recognize 99.9 percent of all objects close by. But put the robot, or the object, in motion, and the readings can become inaccurate. Given a robot or object that moves fast enough, detection may be missed entirely.

Adding a Rotating Turret

With a motorized sensor turret your robot can scan its environment without ever taking a step—or rolling a wheel. One or more sensors are attached to an R/C servo motor, which is commanded to sweep back and forth while the sensors send back data.

Accuracy is somewhat diminished when using an ultrasonic sensor when it's attached to a rotating turret. Movement affects the reading. During any motion you can use the measurements for general proximity detection, but you may wish to momentarily stop the turret (and robot) to get a more accurate distance reading.

Figure 6-4 shows a sensor turret with a Ping ultrasonic ranger. The brackets and mounting hardware are available as a kit from Parallax, or you can make your own.

Refer to Figures 6-5 and 6-6 for a schematic and breadboard view, respectively, of connecting the turret servo to the Arduino.

Figure 6-4   A servo turret, which scans the ultrasonic ranger in a 180-degree arc. It is attached to the front of the Teachbot using simple brackets, which are included in the turret kit.

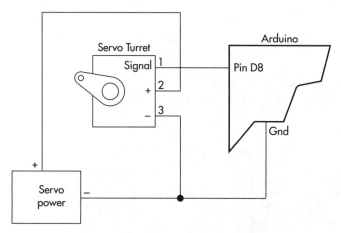

Figure 6-5 Connection schematic for the servo turret. You use the same power supply for this servo as the Teachbot's two main drive motors.

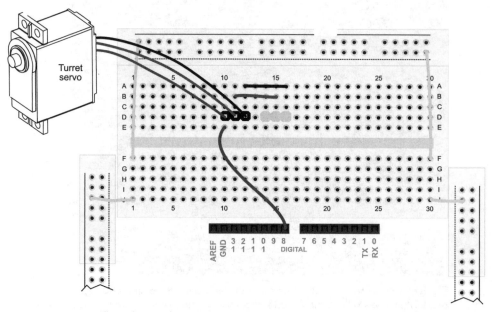

Figure 6-6 Breadboard view of connecting the servo turret to the Arduino. Note that you need to already have the servo wiring in place (see Chapter 4).

Teachbot_TurretScan shows a demo sketch for rotating the turret in small segments of an arc, sweeping from one side to the other. I've intentionally limited the span of the turret to prevent the wires from getting tangled, and from having either sensor deliver inaccurate results because of the wire harness at the rear. As shown in the sketch, the turret has an approximate 250-degree span.

Teachbot_TurretScan

```
#include <Servo.h>

Servo myServo;
const int delayTime = 2000;
const int servoPin = 8;

void setup() {
  myServo.attach(servoPin);
}

void loop() {
  turretL();
  turretR();
  // Rotate turret in steps
  //  Arguments: min-position, max-position,
  //  delay between steps
  turretStep(0, 180, 150);
  turretCenter();
  delay(delayTime);
}

void turretR() {
  myServo.write(0);
  delay(delayTime);
}

void turretL() {
  myServo.write(180);
  delay(delayTime);
}

void turretStep(int minVal, int maxVal, int stepDelay) {
  for(int i=minVal; i<=maxVal; i+=6) {
    myServo.write(i);
    // Read ultrasound here
    delay(stepDelay);
  }
}

void turretCenter() {
  myServo.write(90);
  delay(delayTime);
}
```

As a demonstration, the turret is made to stop periodically every 30 degrees, or six times when going from one extreme to the other in its 180° arc. During each stop you'd take a reading from the ultrasonic sensor. Stopping the turret each time to take a reading isn't a requirement, but it helps improve the accuracy of the measurements.

Adding a Sharp GP2Y0D810 Infrared Detector

Sharp's GP2Y0D805 and GP2Y0D810 sensors provide proximity detection using a beam of infrared light. These are *digital* sensors in that they only provide a no or yes result, depending on whether the object is within a predefined range. The '05 has a set 5-cm proximity range; the '10 has a set 10-cm range. Output is a digital LOW (the object is nearby) or HIGH. These sensors come in a DIP-size package, making them small and easy to mount on smaller robots. But the sensors require some external parts to complete the circuitry. Several online sources, such as Pololu, offer the '05 and '10 sensors on a breakout board for easy use in your projects. Even with the addition of the breakout board, these sensors are roughly half the size of similar infrared sensors, and they're less expensive.

Figure 6-7 shows a Sharp GP2Y0D810 mounted on the "rear" of the Teachbot, acting as a kind of backup alarm. See Figure 6-8 for the simple wiring connection between the Arduino and the sensor. Use a three-wire female-female servo extension to connect between the sensor and the Teachbot's breadboard, as shown in Figure 6-9.

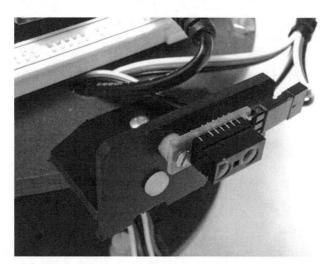

Figure 6-7 A Sharp GP2Y0D810 infrared distance judgment detector mounted to the "back" end of the Teachbot. I used a small plastic bracket and attached the bracket to the top deck of the Teachbot using a single machine screw.

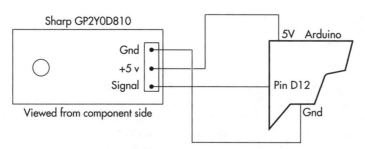

Figure 6-8 Wiring diagram for connecting to the Sharp GP2Y0D810

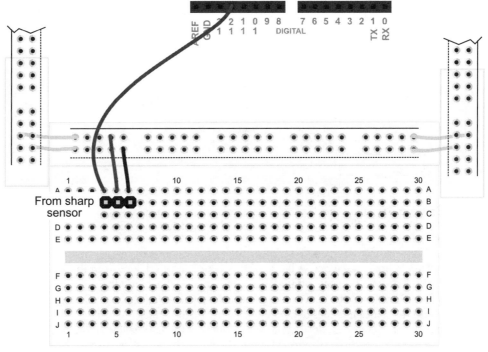

Figure 6-9 Breadboard connection for the Sharp GP2Y0D810 IR sensor

The *Teachbot_Sharp810* sketch provides a simple demonstration of using the GP2Y0D810 sensor to detect when an object is within range. As you can see, the digital LOW/HIGH output of the '810 makes it as easy to interface to the Arduino as an ordinary switch.

Teachbot_Sharp810

```
const int irPin = 12;

void setup() {
  Serial.begin(9600);
}

void loop() {
  Serial.println(digitalRead(irPin), DEC);
  delay(100);
}
```

Adding a Sharp GP2D120 Infrared Detector

An older style, but still available, variation of the Sharp infrared sensors is the GP2D120. Unlike the GP2Y0D805 and GP2Y0D810, the '120 provides a variable voltage analog output. The voltage changes depending on how close an object is to the sensor. These sensors rely on the displacement of reflected light across a linear sensor (see Figure 6-10).

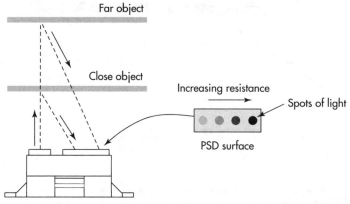

Figure 6-10   The Sharp infrared distance modules rely on the displacement of reflected light across a linear plane to detect variations in distance.

Here's how they work: A beam of modulated infrared light from the sensor illuminates an object. The beam reflects off the object and bounces back into the sensor. The reflected beam is focused onto what's known as a *position sensitive device,* or *PSD.* The PSD has a surface whose resistance changes depending on where light strikes it. As the distance between sensor and object changes, so does the linear position of the light falling on the PSD. Circuitry in the sensor monitors the resistance of the PSD element and calculates the distance based on this resistance.

- The GP2D120 is an example of a *distance measurement* sensor. The voltage output is nonlinear, as shown in Figure 6-11. These detectors work over a span of minimum and maximum distance, usually no closer than a few inches, and up to several feet.
- Conversely, the GP2Y0D805 and GP2Y0D810, detailed in the previous section, are *distance judgment* sensors. They provide a simple digital (LOW/HIGH) signal that represents whether an object is within detection range. That range is set at the factory, and depends on the specific model of the sensor.

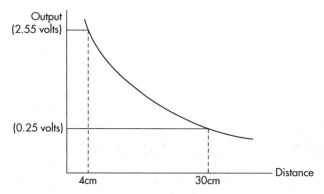

Figure 6-11   The analog output of the Sharp GP2D120 sensor is a nonlinear voltage of between (approximately) 0.25 volts and 2.55 volts.

Sharp has a habit of adding new sensor models and dropping others. The GP2D120 is among a series of sensors that have been discontinued, but as of this writing they are still available through a number of online sources, such as SparkFun and Pololu.

CONNECTING THE GP2D120 TO THE ARDUINO

Distance judgment sensors are ideal for interfacing with simple electronics, as they don't require analog-to-digital conversion (ADC). Since the Arduino is equipped with a multiple-input ADC, we can use either type. The GP2D120 has a range of 4 cm to 30 cm (about 1.5 inches to 12 inches). The distance is reported as a varying voltage, from approximately 0.25 volts (no detection) to 2.55 volts (detection at minimum distance). That's according to the spec sheet, but know that there can be a normal variation of a few tenths of a volt from one sensor to another.

While the GP2D120 is capable of reporting distance with acceptable accuracy, for the Teachbot I've elected to use it as a "multi-zone" proximity detector; that is, instead of hassling with converting its analog voltage to some quasi-precise distance measurement, the GP2D120 will instead simply indicate when an object is within preset zones. The Teachbot relies on a separate ultrasonic rangefinder for accurate distance measuring, and as a secondary near-object detection check.

See Figure 6-12 for a schematic diagram for connecting the GP2D120 to the Arduino. Figure 6-13 has the same circuit, but in breadboard view.

Like many of the Sharp infrared distance sensors, the GP2D120 can introduce electrical noise into your robot's power supply. This can cause false readings in the sensor, or affect other electronics on the Teachbot. You can help filter the noise by adding the *decoupling* capacitors across the +V and ground power connections.

Place the capacitors right at the connection points on the breadboard. Be sure to observe correct polarity of the electrolytic capacitor.

By the way, a decoupling capacitor is an ordinary capacitor. It's simply used in the role of decoupling (really, filtering) the noise between the sensor and the rest of the robot's electronics.

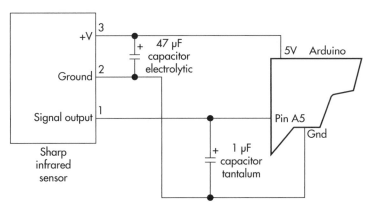

Figure 6-12 Connection schematic for the Sharp GP2D120

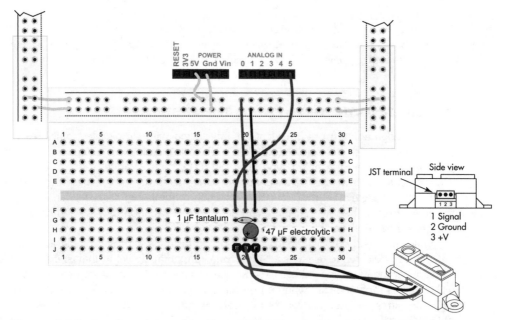

Figure 6-13 Breadboard view of the Sharp GP2D120-to-Arduino connection. Note the pinout and wiring order of the GP2D120 sensor. The wiring diagram relies on existing connections on the breadboard.

The *Teachbot_SharpIR120* sketch demonstrates how to use the GP2D120 with the Arduino. The sketch takes five "samples," each separated by a delay of 55 ms. The five samples are averaged to remove possible incorrect readings due to momentary glitches. Open the Serial Monitor window to observe the actual values reported by the sensor, converted from an analog voltage to 10-bit digital values (0 to 1023) by the Arduino's ADC.

Teachbot_SharpIR120

```
const int irSense = A5;      // Analog pin A5
int distance = 0;

void setup() {
  Serial.begin(9600);      // Use Serial Monitor window
  delay(250);              // Wait 1/4 second to stabilize
}

void loop() {
  Serial.println(irRead(), DEC);
  delay(1000);             // Take one reading every second
}

int irRead() {
  int averaging = 0;
  // Sampling 5 readings from sensor
  for (int i=0; i<5; i++) {
    distance = analogRead(irSense);
```

```
    averaging = averaging + distance;
    delay(55);        // Wait 55 ms between each read
  }
  distance = averaging / 5;        // Average out readings
  return(distance);
}
```

You won't get the full 1024 steps because the sensor does not output a full 5 volts when an object is closest. Minimum values are about 40 to 60; maximum values are in the 625 to 675 range, depending on the sensor and the reflectivity of the object (dark colors tend to produce slightly lower values). Anything outside these ranges can indicate a spurious reading.

Take note of the connection diagram for the GP2D120. Sharp uses a polarized JST connector where the +V power lead is on the outside and ground is in the middle. This is potentially dangerous when the other end of the wiring is nonpolarized, as is often the case with connectors on general-purpose microcontrollers.

Numerous sources, such as Lynxmotion, sell adapter cables that go from the JST locking connector to a standard three-pin 0.100" female header. On many of these cables the wiring order is "corrected" to place the +V power lead in the center. In this arrangement damage is less likely to occur if the connection is flipped. I have shown such a cable in the breadboard wiring view in Figure 6-13. Regardless of whether you use a cable type that rearranges the wiring order, *be absolutely sure to observe correct polarity.* Or poof goes your sensor.

The *Teachbot_SharpIR120* sketch provides a simple operational example only. Due to its length, I've moved the full example showing the GP2D120 combined with the other sensors on the ARB Support Site (see Appendix A). There, you'll see how to use the GP2D120 with the Ping ultrasound and other sensors.

Adding a Power Switch to the Teachbot

I've saved this mini-project for last because it's an optional, but recommended, feature you can add to your Teachbot. As you work with the Teachbot you'll find it gets tedious to keep plugging and unplugging its two battery supplies.

Instead, add a switch so that you can simply flick power on and off when you need it. Because the Teachbot uses two separate power supplies, you need a double-pole, double-throw switch with a center-off position. Wire the switch as shown in Figure 6-14.

- When the switch is in the center position, the Teachbot is completely unpowered.
- When the switch is the "A" position, only the Arduino is powered. The motors don't receive juice. Use this position when you want to experiment without having the motors run.
- When the switch is in the "B" position, both the Arduino and motors are powered. Use this position for normal operation of the Teachbot.

You can see a completed switch, mounted in a homemade bracket, in Figure 6-15. (I used a 1/8" piece of plastic, cut to shape, with holes drilled to mount the switch.)

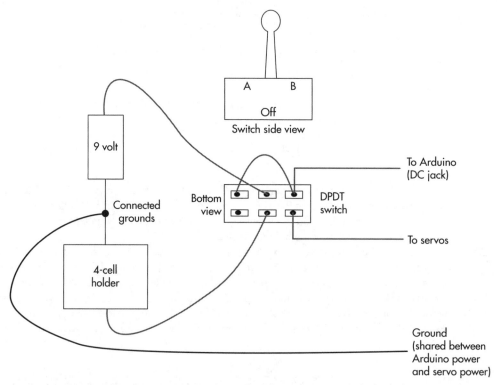

Figure 6-14 How to wire a DPDT (double-pole, double-throw) switch to control both battery supplies of the Teachbot. Be sure the ground connections of the two battery packs are connected at one central point. This can be in the wiring from the battery packs, or on the solderless breadboard.

Figure 6-15 The DPDT switch, attached to a homemade bracket, and mounted on the bottom deck of the Teachbot, for easy access

Note that with most (but not all) switches, the "A" and "B" positions are opposite their terminals on the back of the switch. This is due to the internal construction of the switch. Don't worry if the "A" and "B" of your switch are reversed from the description here. It's all relative.

Review of Teachbot Servo Connections

Lots of things are connected to the Teachbot. Consult Figure 6-16 for an overview of the servos, sensors, and other components attached to the various Arduino digital and analog pins.

As you can see, there are still plenty of free pins available, and space here and there on the two breadboards on top of the Teachbot. Feel free to experiment with your own designs, but watch the power draw of whatever you add. Keep in mind the following:

- Servos should be connected to the servo power pins, exposed on the top breadboard.
- The Arduino and all other sensors and components are powered by a single 9-volt battery, which is capable of delivering about 400 milliamps (mA) of juice. Some types of sensors draw more current, so you'll need to beef up the battery used to supply the Arduino and sensors. A small (AAA-size) six-cell rechargeable battery pack for radio control will deliver 7.2 volts. Fit it alongside the holder for the servo batteries.

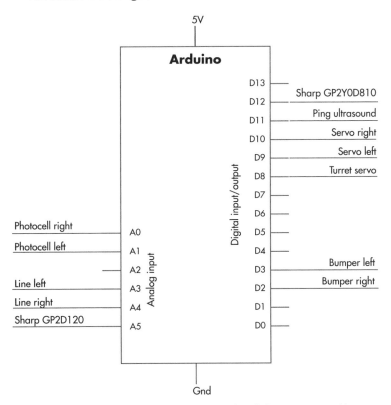

Figure 6-16 "Pin budget" for the Arduino for all the various Teachbot subsystems. There's room to grow.

Making Things

Chapter 7

Making Things: Mechanical

Making and running a modern robot involves three core disciplines: mechanical construction, electronic construction, and programming. Mistakes and shortcomings in any one of these can mean a dead bot. That's not good. We want our bots to operate in tip-top shape.

The best way to ensure carefree operation is to make your robots using solid and dependable construction techniques. In this and the next chapter, you'll learn the basics of just that. In this chapter you'll learn about mechanical construction techniques to help ensure your robot doesn't fall about as it rolls down the hallway.

This book is about desktop robots... those under about 10 to 15 pounds. That means the materials and construction advice is tailored to this smaller scale. Larger robots—the kind that fight other bots or are made to fetch a can of soda from the fridge—require materials and construction better tuned to their weight and size. These topics are outside the domain of this book, but you can find information on building Mongo-sized robots both on the Web and in other books. Refer to the information and sources guides at the ARB Support Site (see Appendix A).

Then in the next chapter, you'll learn about the electronic construction techniques that build a better bot, including ways to simplify wiring things together. As many robots today are "built" by connecting separate pre-made parts together, the next chapter will also help you understand the requirement of correctly interfacing all the pieces.

Some of the information in this and the next chapter is covered to some degree in the construction plans for the various Arduino-based robots in this book. I decided a very minor amount of duplication was better than having you always flip back and forth as you construct your robot. I also wanted to consolidate the bulk of the construction how-to in a single spot, so that if you are already familiar with such things, it won't interfere with your bot-building bliss.

Important Tools for Making Robots

These are the "must-have" tools for any robot workshop. Except for the drill, all are hand-operated, and most are likely ones you already own.

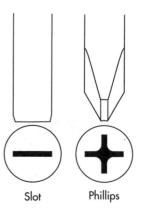
Slot   Phillips

- *Glasses or goggles eye protection.* Always be sure to wear eye protection (see Figure 7-1) to guard against flying debris. Wrap-around style provides the best protection. Use clear lenses for indoors; shaded lenses outdoors in sunlight.
- *Tape measure.* Keep a retractable 6- to 12-foot steel tape measure handy at all times. Graduations in both inches and metric can be helpful, but are not critical.
- *Screwdriver assortment.* Invest in a good set with slot and Phillips head. At a minimum, get #1 and #2 Phillips, and medium- and small-tip slotted drivers. Magnetic tips are handy, but not necessary. Purchase a good set, and test the grips for comfort. Miniature fasteners (those 4-40 and under) require smaller #00 and #0 Phillips screwdrivers.
- *Standard and needle-nose pliers.* A pair each of standard and needle-nose pliers is sufficient for 90 percent of all jobs requiring you to grip parts during assembly. For heavy-duty jobs, purchase a pair of "lineman's" pliers.

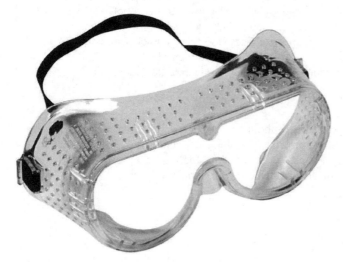

Figure 7-1 Always remember to use adequate eye protection whenever building your robots. The wrap-around type is best.

Figure 7-2 The hacksaw is the common cutting tool in the robot builder's toolbox. Get several blades, with different tooth pitches. Coarser pitches are meant for softer materials, like wood.

- *Wrench, adjustable or box.* Adjustable wrenches let you "dial in" the size of fastener you're using. For smaller bots get one that has a maximum 1/2" opening. If you prefer, you can use a set of box wrenches to match the nut sizes you typically use. As reference, a standard 4-40 nut uses a 1/4" wrench.
- *Hacksaw with blade assortment.* A hacksaw, like the one in Figure 7-2, is the primary cutting tool for small robots. Look for a model that allows quick blade changes, yet holds the blade securely. Purchase an assortment of carbide-tipped blades in 18 and 24 teeth-per-inch (tpi)—the finer tooth blades are for harder materials.
- *Electric drill and bits.* Make holes with an electric motorized drill. For greatest flexibility get one that's reversible with speed control. A 1/4" or 3/8" chuck (the part that clamps down on the drill bit) is preferred. Along with the drill, purchase a set of steel twist bits. For users in the United States, these typically come in a set of fractionally sized bits, starting at about 1/16". For small robots there's only occasional need for bits over 1/4", so get these as needed. The better bits come with coatings to help keep them sharp. If all you're doing is drilling into plastic, the expensive coated bits aren't necessary.
- *Rasps, files, and sandpaper.* Rasps are used to remove large amounts of material from the work, while files are for smoothing out. Use sandpaper to finish edges and surfaces, especially if you're using wood. Rasps come in several thousand combinations of sizes and styles. It's best to get a small general-purpose set as a starter.

 Files are available for either woodworking or metalworking; they're different enough that you'll want to pick one over the other, depending on the material of your robot. Hard plastics will work with most metalworking files; soft plastics with woodworking files.

Robot Construction Materials

Robots can be constructed from many materials. The most common materials used in constructing robot bases and other components are wood, plastic, aluminum, and lightweight foam board.

ROBOTS MADE OF WOOD

Wood is easy to work with, can be sanded and sawed to any shape, doesn't conduct electricity (avoids short circuits) unless wet, and is available everywhere. Disadvantage: Ordinary construction woods are not recommended; use the more dense (and expensive) multi-ply wood for model airplane and sailboat construction.

Wood plys alternating direction of grain

Figure 7-3 Plywood is made from alternating the direction of grain of multiple layers of wood. The layers are bonded together with strong adhesive.

Wood can be broadly categorized as hardwood or softwood. The difference is not the hardness of the wood, but the kind of tree the wood is from. *Hardwood* is produced from trees that bear and lose leaves (deciduous). *Softwood* trees bear needles (coniferous) or do not undergo seasonal change (nondeciduous). In general, deciduous trees produce harder and denser woods, but this is not always the case. A common hardwood that's very light and soft is balsa.

Wood has a grain, and its strength is in the direction of the grain. One way to make wood stronger is to layer many thin sheets together, where the grain of each sheet alternates direction. This is the concept behind plywood, as shown in Figure 7-3. The best plywoods for robot building are the ones made for making model airplanes and making crafts. They come in various thicknesses, determined by the number of plys. A 6mm (about 1/4") aircraft plywood might have from 5 to 8 plys.

ROBOTS MADE OF PLASTIC

Pound for pound, plastic has more strength than many metals, yet is easier to work with. You can cut it, shape it, drill it, even glue it. Effective use of many types of plastics requires some special tools. Some unique plastics might be harder to get unless you live near a well-stocked plastic specialty store; mail order is a great alternative.

The plastics family is *huge,* and encompasses thousands of materials. Only a few are suited for amateur and educational robotics, due to cost and other factors. These include ABS, acrylic, polycarbonate, and PVC.

ABS Plastic

ABS stands for acrylonitrile butadiene styrene. The most common non-robot application for ABS is sewer and waste-water plumbing systems, but the plastic is also available in sheets (and other forms) from plastic specialty outlets. It is tough and hard and yet relatively easy to cut and drill. ABS comes in rods, sheets, and pipes, and also happens to be the plastic used to make LEGO pieces.

ABS at a glance:

- One of a number of "engineering plastics" because of its machinability, yet is also tough and impact resistant
- Typical commercial uses: Impact-resistant, commonly used in water waste pipes and pipe fittings, electronic housings, auto parts, and automotive trim
- Best robotic uses: Base material, small parts cut to size and shape
- Workability: Easy to drill and cut

Acrylic Plastic

Acrylic is clear and strong, and is the mainstay of the decorative plastics industry. It can be easily scratched, but if the scratches aren't too deep, you can rub them out with a cloth or polishing wheel. Acrylic is somewhat tough to cut without cracking, and requires careful drilling. The material mostly comes in sheets, but also is available in extruded tubing and rods.

Acrylic at a glance:

- Chemically known as *polymethyl methacrylate,* or PMMA
- Plexiglas (U.S.) and Perspex (UK) are common trade names
- Should be avoided in any application with repeated impact or stress, as cracks may otherwise occur
- Can be cut to shape with a desktop laser cutter
- Typical commercial uses: Display materials, works of art
- Best robot uses: Base material, small parts cut to size and shape
- Workability: Difficult to cut and drill without cracking

Polycarbonate Plastic

Polycarbonate is more durable than acrylic and is far more resistant to breakage. Comes in rods, sheets, and tubing. A common inexpensive window glazing material, polycarbonates are hard to cut and drill.

Polycarbonate at a glance:

- Chemical name: polybisphenol-a carbonate
- Optically clear with some cloudiness common, with very high tensile strength
- Best known by the trade names Lexan, Hyzod, and Tuffak
- Can be cut to shape with a desktop laser cutter, but the edges can look "scorched"
- Typical commercial uses: "Unbreakable" window replacement, machine guards
- Best robotic uses: Base material
- Workability: Difficult to cut and drill, but not as prone to cracking as acrylics

PVC Plastic

PVC (polyvinyl chloride) is an extremely versatile plastic best known as the material used in fresh water plumbing and outdoor plastic patio furniture. Though often processed with white pigment, PVC is actually clear and softens in relatively low heat. PVC is easy to cut and drill using ordinary shop tools (see Figure 7-4). Besides plumbing fixtures and pipes, PVC is supplied in film, sheet, rod, and tubing.

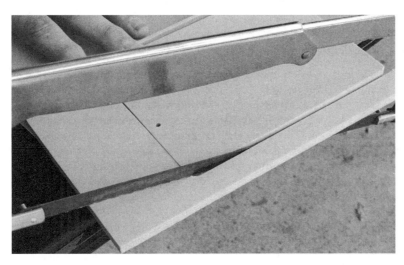

Figure 7-4 Expanded (so-called "foam") PVC is super easy to cut with any power or hand saw.

As robot builders, we're after the rigid expanded PVC sheets used to make signs—sign-makers refer to this raw material as *substrate*. It's available in a variety of sizes and thicknesses, in a rainbow of colors: blue, red, orange, tan, black, brown, yellow, and many others.

Rigid expanded PVC goes by many trade names, such as Sintra, Celtec, Komatex, Trovicel, and Versacel, but it's probably easiest if you just ask for it by its generic "expanded PVC" or "foamed PVC" moniker. Sheets are commonly available in any of several millimeter sizes. Of the more common thicknesses:

Metric Sizing	Approximate Fractional Size
3mm	1/8"
6mm	1/4"
10mm	13/32"

Though expanded PVC is often referred to as "foamed," there is nothing particularly foamy about it. The plastic is not spongy or soft. It's a hard plastic that happens to have a somewhat foam-like cross-section appearance when cut. This appearance is caused by the expansion process during manufacture.

PVC at a glance:

- Used with varying amounts of plasticizer; the more plasticizer that is added, the softer the material
- Material is produced as extrusions, films, granules, and pastes
- Can be combined with other plastics for multi-property benefits (example: ABS/PVC)
- Comes in "expanded" form: light and strong
- Should not be cut with a desktop laser cutter
- Typical commercial uses: Everything, even your credit cards
- Best robot uses: Base material, small parts cut to size and shape
- Workability: Easy to drill and cut

ROBOTS MADE OF ALUMINUM

Before the "age of plastic," metal was the mainstay of the construction material world. Of all the metals available, aluminum is probably the best robot building material for medium and large machines because it is exceptionally strong for its weight. Aluminum is easy to cut and bend using ordinary shop tools. It is commonly available in long lengths of various shapes, but it is somewhat pricey. Other metals, like steel, tin, and brass, are usually more expensive, heavier, and/or harder to work with than aluminum.

You'll find aluminum and other common metals for robot building at the following local sources. If these don't provide the materials you need, try a Google.com search to locate mail order suppliers of the metals you want:

- *Hardware stores* carry some aluminum sheets, as well as various angle brackets (Figure 7-5), straps, and rods made from *extruded* (pushed out of a specially shaped orifice) aluminum.

Figure 7-5 Many hardware and home improvement stores stock various sizes and shapes of aluminum tubing, angles, and strips. Though extra work to cut out, these can make for good robot parts.

- *Hobby stores* sell aluminum, as well as brass and copper, in small sheets, rods, tubes, and strips. A common brand sold by stores in North America is K&S Engineering; you can review this company's product line at *www.ksmetals.com.*
- *Metal supply shops* that cater to welders are open to the public. Many sell stock in large pieces, which you can have cut so you can get it home in your car.
- *Restaurant supply stores,* most of which are open to the public, sell many aluminum and stainless materials. Look for spun bowls, cookie and baking sheets, unusually shaped utensils, strainers, even salt and pepper shakers.

ROBOTS MADE OF FOAM BOARD AND CARDBOARD

Art supply stores stock "foam board," also called "Foamcore," a popular brand. Foam board is a sandwich of paper or plastic glued to both sides of a layer of compressed foam. The material comes in sizes from 1/8" to over 1/2", with 1/4" being fairly common. Foam board is especially well suited for small robots where light weight is important, and also for making quick prototype models.

 Paper-laminated foam board can be cut with a sharp knife; plastic-laminated foam board should be cut with a saw. Remember! Exercise extreme care when using a knife or other sharp tool.

Robots made from foam board aren't as strong as those constructed from stronger and more rigid materials, but sometimes all you need is a temporary working platform to test your ideas. Or you may need to whip up something fast and don't have the time to cut your robot parts from more traditional materials, such as wood or plastic.

Another option for low-cost and low-weight robots is cardboard, the kind used to make shipping boxes. Common cardboard consists of two thin sheets (liner boards) of paper sandwiched to a corrugated inner core. The thickness and material used for the inner core determine the overall rigidity of the cardboard.

Cardboard for shipping boxes is too lightweight for most robot projects. A better kind of cardboard is called *honeycomb board.* This is a special kind of cardboard that uses a thick honeycomb core. Thickness can vary from 1/2 inch to several inches. Honeycomb board is available at many art supply stores, and is used in model making. For robotics, it can be sandwiched to a thin carrier, such as plywood or plastic, or even metal. The carrier need not be thick; the strength of the sandwich material will come from the cardboard.

Compared to both cardboard and foam core, honeycomb board is quite expensive. However, it has good structural characteristics. When used properly, honeycomb board can hold over 100 pounds.

Putting Things Together

How the parts of your robot are connected together is just as important as the parts themselves. Assembly techniques are those that fasten or bind the robot parts together. Many robots use multiple assembly techniques.

Among the most common assembly techniques are:

Construction Technique	Applicable to	Not Applicable to
Snap-together parts	Small toy-based robots (e.g., LEGO or K'NEX)	Designs that require permanence, unless parts are glued
Adhesive tape (electrical tape, duct tape, etc.), double-sided foam tape, standard strength hook and loop (e.g., Velcro)	Test designs; temporary constructions; lightweight materials and components; heavy-duty hook-and-loop (e.g., industrial Velcro) applicable for larger weights and stresses	Long-term constructions; material or component weights of over a few ounces
Glues	Dependent on adhesive, can be used on robots of up to several pounds	Heavier metal robots; components that may receive heavy, sudden shock
Machine screw fasteners	General-purpose robots, but size and weight limitations apply	Very high strength and impact applications (e.g., heavy combat robots); outdoor use unless stainless steel fasteners are used; larger/heavier robots require large bolts, which can add to weight (welding is advisable as an alternative)
Rivets	Same as machine screw fasteners, but rivets provide weight savings	Constructions that may need disassembly; high sheer strength applications
Soldering and brazing	Small to medium size and weight robots	Very large or heavy robots where considerable impact shock may occur
Welding	Large and/or heavy robots, particularly those used in rugged environments	Small or low-cost robots

Let's take a closer look at the most popular approaches to materials assembly.

ADHESIVE TAPES

Adhesive tape is a broad family of products that have an adhesive on one or both sides of the tape. While tape adhesive may be strong, few products are meant to be permanent. This may be exactly what you're after, but if you're not, tape that unravels, or causes parts to fall off, can be a source of frustration.

- Double-sided foam tape that has a spongy middle and adhesive on both sides is an important part of many small robots. You can get it just about anywhere, including craft, department, and hardware stores. The permanence of the adhesive varies between products; some double-sided foam tapes are engineered with an adhesive that never fully cures. It stays gummy so that the tape can be more readily removed from walls.
- Duct tape is known for its highly sticky adhesive. It's made to seal air ducts for ventilation systems—where it gets its name. Both the tape itself, and the adhesive, are flexible. This makes duct tape ill suited for mounting motors or other components that must be kept aligned. Though some duct tape is plastic, much of it is made of paper, and is not waterproof.
- Gummy transfer tape provides "stickup" without the tape. This material is engineered to leave its sticky residue, but no tape. It's meant for such jobs as electronics production, where workers apply small dabs of sticky substances to hold down wires and components. Transfer tape is available from art supply stores, adhesive specialists, and electronics production supply outfits.

 Some gummy transfer tapes work best when used with a specially made dispenser. The dispenser can add to the cost, so if you only plan on using a little bit of transfer tape, either get the kind that already comes with the dispenser, or opt for a brand that offers the convenience of peel-and-stick.

- Self-cling stretch film is made to stick to itself. It's good for holding wires and other parts together on the robot. To use the plastic, cut into usable pieces, and clean with denatured alcohol. Apply the tape by rolling it around itself. The cling will not be particularly strong, but should be adequate for such tasks as bundling wires and temporarily holding small parts together for subsequent gluing.

HOOK-AND-LOOP

Hook-and-loop is two pieces of fabric where the parts stick to themselves. You can use it for mounting small or lightweight parts to the body of your robot, or to bundle wires and other components to keep them neat. Velcro is a trade name for a kind of hook-and-loop fabric fastener.

Among the most useful hook-and-loop products is the continuous strip, where you can cut what you need to length. The strip comes in packages of one foot to several yards, in any of a number of widths—1/2" and 1" wide are the most common. The strips come with a peel-off adhesive backing that allows you to directly apply the hook and the loop to the parts you wish to fasten, like that in Figure 7-6. If the adhesive is not strong enough (which is sometimes the case), you can reinforce the material with a heavy-duty epoxy, screws, or staples.

Figure 7-6 Hook-and-loop fabric (Velcro) can be used to attach parts of your robot together. Get the heavy-duty or industrial version for heavier and larger robot parts.

PLASTIC TIES

Plastic ties are intended to hold bundles of wire and other loose items. They can also be used to hold things to your robot. The tie is composed of a ratcheted strip and a locking mechanism. Loop the strip into the mechanism, and pull the strip through. The locking mechanism is one-way: you can tighten the strip, but can't loosen it (applies to most plastic ties; some have a releasable lock).

Plastic ties are made of nylon and are very strong and durable. They're available in a variety of lengths, starting at 100mm (a little under 4") to well over 12". Save the larger ties for the heavy-duty jobs. For most applications, the 4 to 6" lengths will work fine. You can anchor the tie into a hole drilled for the purpose, or use one of several mounts specifically designed for use with plastic ties. I prefer mounts designed for use with hardware fasteners. Drill a hole for a 4-40 or 6-32 machine screw, and secure the mount with a screw and nut.

CONSTRUCTION WITH GLUE

Glue is important stuff in any building endeavor. Glue is used to simplify construction, and can reduce the weight of the finished piece over rivets or fasteners. When properly applied, glued joints can be just as strong, and in some cases stronger, than the base material itself. Parts of airplanes are glued together rather than riveted or fastened, both as a cost-savings technique and to make the plane lighter.

There are hundreds of different glues available, and many are made for bonding specific materials to another. For instance, a wood glue is best for bonding wood to wood (imagine that!). As availability of glues varies from one locale to another, and their use is fairly self-explanatory, your best bet is to visit your local hardware or home improvement store and browse through their selection. The labels on each glue tube or bottle will tell you what it's for.

CONSTRUCTION WITH MECHANICAL FASTENERS

Mechanical fasteners include nuts, screws, bolts, and other hardware used to hold pieces together. Fasteners are favored because they are cheap, easy to get, and unlike adhesives or even welding, don't require that you exactly match the materials to be joined.

Fasteners are the most elementary of all hardware. There are dozens of fastener types, and we'll concentrate just on those that are most practical for amateur robotics. These are machine screws, nuts, and washers (see Figure 7-7).

- *Machine screws* are designed for fastening together the parts of machinery, hence the name. Unlike a wood or metal screw, machine screws do not have a pointed end. The machine screw is designed to be secured into a nut or other threaded retainer. If the material being joined is threaded, the screw can be secured directly.
- *Nuts* are used with machine screws. The most common is the hex nut, so called because the nut has six sides. The nut is fastened using a wrench, pliers, or hex nut driver. Also handy in many robotics applications are *locking nuts,* which are like standard hex nuts, but with a nylon plastic insert. The nylon helps prevent the screw from working itself loose.
- *Washers* act to spread out the compression force of a screw head or nut. Under load and without a washer, damage may occur because of the small contact area of a machine screw head or nut. The washer doubles or even triples the surface area, spreading out the force. Washers are available in diameters to complement the size of machine screw (or bolt for larger sizes). Variations on the washer theme include *tooth* (internal or external) and *split lock washers;* these provide a locking action to help prevent the fastener from coming loose.

Understanding Fastener Sizes

Fasteners are available in common sizes, either in metric or standard.

Standard may also be referred to as SAE, English, American, fractional, inch, unified, and other names. For the sake of simplicity, I'll refer to it just as standard, even if the term is not always 100 percent applicable to every nuance of fastener science.

Figure 7-7 The three most common fastener types: nut, screw, and washer. Each of these comes in many subtypes, though they're not often used in building small desktop robots.

Standard-size fasteners are denoted by both their diameter and the number of threads per inch. For example, a fastener with a thread size of 6-32 (also shown as 6/32) has a diameter referred to as #6, with 32 threads per inch (also called *pitch*). Diameters under 1/4" are indicated as a # (number) size; diameters 1/4" and larger can be denoted by number, but are more commonly indicated as a fractional measurement—3/8", 7/16", and so on.

The pitch (number of threads per inch) can be either *coarse* or *fine* for standard fasteners. Therefore, not all #6 fasteners have 32 threads to the inch. Note that sizes #14 and above are more commonly referred to using a fractional measuring system.

Coarse threads are by far the most common at hardware stores, though #10 screws are routinely available either coarse or fine. This can actually be a problem, because it's easy to buy 10-32 nuts and not notice you have 10-24 machine screws.

As you might guess, metric fasteners don't use the same sizing nomenclature as their standard cousins. Screw sizes and pitches are defined by diameter, the thread pitch (number of threads per millimeter), followed by length—all in millimeters. For example:

M2-0.40 × 5mm

means the screw is 2mm in diameter, has a pitch of 0.40 threads per millimeter, and has a length of 5mm. Metric screws use a normalized thread pitch, so the pitch size may be omitted, as in:

M2 × 5mm

FASTENER HEAD AND DRIVER STYLES

Machine screws differ in both the type of head they use and the driver (as in screwdriver) they're designed to work with. The most common are (see Figure 7-8):

- *Pan.* This is a good general-purpose fastener. However, the head is fairly shallow, so it provides less grip for the driver.

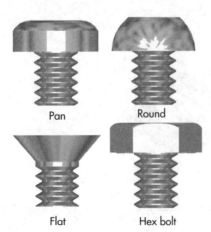

Pan

Round

Flat

Hex bolt

Figure 7-8 Four common screw head types: pan, round, flat, and hex

Slotted   Phillips

Combination   Socket

Figure 7-9   Four common screw driver types: slotted (or flat), Phillips, combination, and hex socket. Variations on the hex socket type include heads that require specialty tools. Among the more common are Torx and square.

- *Round.* The taller head protrudes more than pan head, so it provides greater depth for the driver. Good for higher torque applications when it doesn't matter if the head sticks out.
- *Flat (or countersunk).* Used when the head must be flush with the material's surface. Requires a countersunk hole.
- *Hex bolt.* Uses no slot, and requires a wrench to tighten. It's intended for the highest torque applications.

The type of tool you use to tighten the screw is defined by its driver, shown in Figure 7-9. Here are the ones you'll encounter most often. In all cases, you must match the tool to the driver size required by the fastener.

- *Slotted.* Made for general fastening and low torque drive, as the screwdriver may slip from the slot.
- *Phillips.* Cross-point drive resists drive slippage, but the head is easily stripped out when using an improperly sized driver.
- *Combination.* Both slotted and Phillips, in case you don't have the right screwdriver handy.
- *Socket.* Uses a six-sided (hex or Allen) drive in the form of an angled wrench or screwdriver. The rim of the head may also be knurled in order to assist in hand-tightening.

The Construction Process

Building the typical robot requires only basic handiwork skills, principally drilling, cutting, and sanding (or other finishing step).

THE SCIENCE OF GOOD DRILLING

Regardless of the material you use—whether it be wood, plastic, aluminum, or whatever—keep in mind the following points.

Aligned Holes

Drilling with any hand-held tool will naturally produce a certain amount of error. Even the most skilled worker cannot always drill a hole that is at exact right angles to the surface of the material. When precisely aligned holes are a must, use a drill press or drill-alignment jig. The latter can be found at better hardware and tool stores that stock specialized carpentry accessories.

Clamping and Other Work Hold-downs

When drilling with power tools, care must be exercised to hold the part in a clamp or vise. Without appropriate hold-down methods, the drill may "dog" into the material, causing it to spin with the bit. Even a piece of wood turning at a few thousand rpm (revolutions per minute) can cause serious injury.

Hold-down clamps and vises come in various shapes, sizes, and styles. There is no one type that works for every occasion. Spring-loaded clamps (they look like giant tweezers) are useful for very small parts, while C-clamps are handy for larger chunks of material. A vise is required when drilling small parts of any type with a drill press.

Setting the Right Drill Speed

Most drill motors lack a means to directly measure the speed of the tool, so the rpm should be considered an approximation only. As you gain experience you can go by the sound of the tool; until then use the following table as a guide. If you know the "full speed" of your particular tool is 6500 rpm, you can recognize approximate half and quarter speeds by listening to the sound of the motor.

	Wood	**Aluminum**	**Plastic**
General	Wood is readily drilled using a motorized drill, either hand-held or drill press. Speed is dependent on the size of the bit and the density of the wood. Following are general speed recommendations (all are in rpm, revolutions per minute). Larger than 1/8": 2100 rpm 1/8" to 1/16": 4500 rpm Smaller than 1/16": 6200 rpm	Metals should be drilled using a motorized drill. Small parts are more readily drilled using a drill press. Following are general speed recommendations for aluminum and other soft metals (all are in rpm). For harder metals, reduce speed by 50 to 75 percent. Larger than 1/8": 1000 rpm 1/8" to 1/16": 1500 rpm Smaller than 1/16": 2000 rpm	For soft plastics (e.g., PVC), speed settings same as for wood. For harder plastics (acrylic, polycarbonate), reduce drill speed by 30 to 40 percent.
Bits	Wood bits should be ground to 118°. For cutting all but very dense hardwoods (e.g., oak), standard carbon twist drills are more than adequate.	For most metals, bits should be ground to 118°. For harder metals, use 135° (split point) bits. For longer life, consider titanium- and cobalt-coated bits.	Use wood bits for soft plastics. For hard plastics, use a pointed bit designed for acrylic, polycarbonate, and similar plastics.
Cooling	Air cooling is sufficient. If wood is very hard and thick, pause every 30 seconds to allow bit to cool down.	Use cutting oil for heavy-gauge metals. Always avoid excessive heat, as this will dull the bit.	Air cooling is sufficient, but if plastic remelts into the hole, consider use of small amount of water.

THE ART OF GOOD SAWING

Wood, metal, and plastic can be cut using hand-operated or power tools. For all but the lightest materials, however, you will find that power tools make short work of the job.

For hand tools, practical choices include:

- For wood, a *backsaw*; for metal and plastic, a *hacksaw*. The backsaw will need periodic resharpening to keep it in top form. For the hacksaw, you need only replace the blades when they become dull.
- A *coping saw* allows you cut tight radius corners in wood, metal, and plastic. A coping saw is similar to the hacksaw, except the blades are smaller. Replace the blade when it's dull.
- A *razor saw* is used with thin woods and plastics. Its shape is like that of a backsaw, but much smaller. You can find razor saws at the hobby store.

For power tools, practical choices are:

- *Jigsaw* for wood (Figure 7-10).
- *Circular saws* and *table saws* are useful for cutting long, straight cuts in wood and metal. Be sure to use the proper blade, or else damage to the material and/or blade could result.
- A *bandsaw* can be useful for cutting small shapes in wood, metal, or plastic. The bandsaw is useful when cutting plastic because the long length of the blade (usually 40" or more) stays relatively cool as it goes through the machine. This reduces remelting of the plastic. The better band saws have a speed control.
- A *circular miter saw* is useful when cutting aluminum channel and bar stock. The alternative is a hacksaw, a miter block, and sweat.
- A *scroll saw* can cut very intricate shapes in wood and metal and some plastics. The saw cuts by reciprocating action, so the blade can become quite warm when used with metal and plastic. Use a low cutting speed with both metal and plastic to reduce heating.

Figure 7-10   A motorized jigsaw (also called a saber saw) is a handy tool when you need to do a lot of cutting in softer materials such as wood or plastic. Select a blade for the type of material you're using. Harder plastics require a finer-tooth blade.

For best results select a cutting blade and cutting speed to suit the material you are using. Some generalities:

	Wood	**Aluminum**	**Plastic**
General	For hacksaw, band saw, or scroll saw, use medium pitch blade. For motorized tools, set at highest speed.	For motorized tools, reduce speed to 25 to 50 percent.	Set cutting speed to 50 to 70 percent.
Blade	Match the blade with the thickness and grain of the material. Circular saw blades are often classified by their use (e.g., crosscut); use this as a guide.	As a general rule, between three and five teeth should engage the metal. Use an abrasive cut-off tool for heavy-gauge ferrous metals.	Circular saw: If possible, use a "non-melt" blade made for plastics; if not available, a high-quality plywood blade will suffice. The wider the kerf in relation to the thickness of the blade, the better (this avoids remelting). Hack saw, band saw, or scroll saw: Use an intermediate pitch blade (18 to 24 teeth per inch). Use a wide-kerf blade.
Cooling	Air cooling is sufficient.	Use cutting oil or wax for heavy-gauge metals.	Air cooling is usually sufficient. If remelting occurs, direct 50 to 75 psi air from a compressor over cutting area.

Chapter 8

Making Things: Electronic

Building a robot means connecting together various electronic stuff. Maybe it's just plugging in wires between a robot controller and a pair of motors. Or it might be a complete circuit board that you made yourself.

Apart from the mechanical construction, which you learned about in the previous chapter, building a robot requires at least some electronic making. In this chapter you'll learn about constructing the electronic subparts of your robot—including using solderless breadboards, tools you need to get the job done, and designing your own circuit boards—along with helpful tips and tricks to make the work go faster, and the result last longer.

Also in this chapter you'll discover the fundamentals of interfacing one kind of circuit to another. Why is this important? Many robots today are built by connecting separate pre-made parts together. These subsystems add to the core Arduino to complete your robot creation. As these subsystems are not a part of the Arduino, and may not even be intended for use with it, you must be sure all the pieces are correctly interfaced—that is, their electrical specifications match.

Using Solderless Breadboards

The solderless breadboard is one of those inventions that just might have changed the course of world history. Okay, maybe that's a bit much. At the very least they made it easier to design and test electronic circuits. You literally plug in the various parts of the circuit, then attach some wires to complete the connections. You can easily change the value of components to see if another one works better.

Many of the projects in this book rely on one or more solderless breadboards, like the kind in Figure 8-1. It's a central place for building circuitry and joining the various electronic parts of the robot. For instance, the Teachbot in Chapters 3 through 6 uses several small breadboards placed around an Arduino.

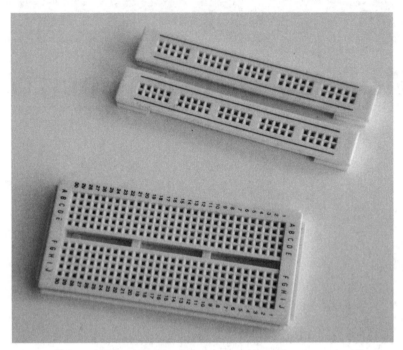

Figure 8-1 Solderless breadboards provide a convenient way to build and test circuits, without having to solder wires or components.

Because solderless breadboards have become so ubiquitous in all things electronic and robotic, they require little instruction on how to use them. So I'll just cover the main topics of interest.

INSIDE THE SOLDERLESS BREADBOARD

Solderless breadboards consist of holes with shared internal contacts. These contacts are spaced one-tenth of an inch apart, which also matches most integrated circuits, transistors, and discrete components like capacitors and resistors. The holes are sized to accept 22-gauge solid conductor wire, though in a pinch you can also use 20 or 24 gauge. On each column you can connect together as many as five wires or components.

Thicker wires may cause the metal contacts inside the breadboard to splay out, making the opening larger. This can cause intermittent connections. Avoid the use of thicker wires and component leads, especially for a breadboard you plan to keep reusing.

Likewise, thinner gauge wires may not make adequate contact, causing disruptions in the operation of your circuit. If the wire is very thin, consider soldering on a thicker extension.

Stranded hookup wire is generally not recommended for use with solderless breadboards, unless you tin the wire ends to keep the strands from breaking off from the group. Tinning means to apply a light amount of solder to the wire. Be sure there are no globs of solder on the tip of the wire that could cause damage to the internal contacts of the breadboard.

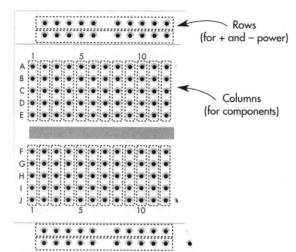

Figure 8-2 The internal construction of a solderless breadboard uses springy metal contacts arranged in columns and rows.

In addition to the column connections, many solderless breadboards have long rows on the top and bottom; see Figure 8-2 for an internal view of the typical breadboard. These rows provide common tie points for positive and negative power supply.

The actual size and layout of solderless breadboards differ. On the small end of the scale are those only a few inches square. These contain 170 contact points and will fit atop most any robot you build. They're great for small and simple circuits containing just a handful of small components. For larger projects, opt for breadboards with 350 to 400 contact points. These require a bit more room on your robot for mounting, but provide extra space for bigger circuits.

If you're using the breadboard directly on your robot, you'll want to mount it to keep it in place. The easiest method is to use double-sided foam tape or Velcro hook-and-loop material. The latter allows you to peel the board off to work on it, then stick it back on when it's done.

 Some solderless breadboards have mounting holes in the corner. Secure the board to your robot using 4-40 flat head machine screws and nuts.

HOW TO MAKE SOLDERLESS BREADBOARD CIRCUITS LAST LONGER

While solderless breadboards are not intended for permanent circuits, they can often be used as such if you employ a few safeguards:

- Use a new board whenever possible. The spring contacts in the board can stretch or wear out, causing intermittent connection problems.
- Cut all component leads to length, and mount components flush against the surface of the board. Avoid parts that are just stuck here and there, and likely to fall out or snag on something.
- Use small rubber bands or 4" cable ties to keep wires in place.

● For maximum longevity, test the circuit for proper operation, then seat the components by applying a very small dab of SuperGlue to the body of the part. Avoid having the glue ooze down into the holes. Obviously, once glued into place the breadboard will not be reusable for another project.

SOLDERLESS BREADBOARD TIPS AND TRICKS

Here are some tips for solderless breadboard success:

Use pliers when working with small parts and wires. Use small needle-nose pliers to plug the end of the wire or component lead into the contact hole. If the wire is short, use the pliers to gently pull it out of the hole when you are done with the breadboard.

Never expose a breadboard to high heat, or the plastic could be permanently damaged. ICs and other components that become very hot might melt the plastic underneath them. Check the temperature of all components while the circuit is under power.

Avoid building a bird's nest where the connection wires are routed haphazardly. The jumble of wires makes it harder to figure out what's wrong with the circuit. And it greatly increases the chance of mistakes.

Solderless breadboards are designed only for low-voltage DC circuits. They are not designed, nor are they safe, to carry AC house current.

Using Shields

Arduino shields are a great way of creating a robot using the building block concept. Shields are designed to extend the functionality of the Arduino, without requiring any (or at least very little) special wiring or additional components. The shield simply plugs on top of the Arduino board, like the one in Figure 8-3.

Figure 8-3 Shields allow you to nearly instantly expand the Arduino, simply by plugging in an auxiliary board.

Pins along the bottom of the shield plug into the I/O sockets of the Arduino, providing a solid and dependable connection. Many shields likewise have their own I/O sockets on the top, so that you can stack two or more shields together. These shield sandwiches are in fact often referred to as *stacks*.

While shields make it convenient to extend the Arduino, their very design often requires tradeoffs and compromises if you want to mix and match several shields in one project. Here are things to keep in mind when working with shields—either singly or in multiple—when building your Arduino-based robots.

I/O CONFLICT

By their nature, shields use specific Arduino I/O pins. If any two shields require the same pins, the I/O conflict can cause anything from invalid data reaching the Arduino to downright damage to the Arduino and any of the shields you're using. So be sure to study the pin-out diagrams of every shield you use. See the section "Dealing with Overlaps in Shield Pin Assignments" if any pins conflict with one another.

Some pin assignments are meant to be reused by multiple devices. This applies to both the I2C data and clock pins (A4 and A5 on the Arduino Uno), and the SPI serial communications pins (pins 10 through 13 on the Uno). Multiple shields can tap into these pins when they are used for their respective communication applications, as long as the appropriate electrical specifications are met.

EXCESSIVE CURRENT DRAW

A few shields, like those for sending and receiving cell phone text messages, can consume copious amounts of power. This power usually comes from the 5V I/O connection along the bottom of the Arduino. This connection in turn is fed by the Arduino's main voltage regulator; on most Arduino boards this regulator is limited to supplying about 1 amp of current.

Adding together too many shields that each demand heavy current can cause erratic behavior, power failure, or an overheated (and possibly damaged) Arduino voltage regulator. Count up the worst-case power requirements of all shields and other connected components—don't forget the LEDs; each one can suck up 10 to 30 milliamps! Make sure you're below 1 amp.

PHYSICAL AND MECHANICAL CONSTRAINTS

Shields with lots of hardware attached to them may inhibit easy stacking. The hardware gets in the way of other shields. This is particularly true of shields equipped with bulky antennas, or those intended for solderless prototyping. Any oversized shield should be stacked on top.

DEALING WITH OVERLAPS IN SHIELD PIN ASSIGNMENTS

If pin assignments between your shields overlap, you'll need to do some rewiring or reconfiguration in order to connect them to your Arduino. Here are several ways of getting your shields to cooperate with one another:

● *Choose a different shield that avoids the conflict.* This is the no-brainer. Look around for another shield that provides the same functionality, but uses different pin assignments.

- *Reconfigure the shield if it's adaptable.* Some shields use jumpers or cuttable solder traces or pads that allow you to change the pins they use to connect to the Arduino. When using cuttable traces, use a sharp hobby knife to cut (break) the trace of the pin you want to protect. You then make a solder bridge to the new pin. Be sure to modify the pin assignments in the sketch to reflect the change. If the sketch calls for using pin D3, for example, and you've switched it to D6, you must update the sketch accordingly.
- *Mount the shield offboard and wire with jumpers.* When space allows, attach the problem shield someplace else on your robot, and use jumper wires to connect it to your Arduino. You can easily select any pins this way.

SPI communications require a separate *slave select* (SS) pin for every SPI device. Alas, most shields are hard-wired to use the same pin (pin 10) for slave select. This makes it incompatible with other SPI shields you may have. You'll need to use one of the options noted here to connect the slave select of the shield to another pin on your Arduino. Your sketch then needs to be told which pin is used for slave select for that device. Refer to the Arduino SPI library online documentation for how to set a different slave select pin.

SHIELD COMPATIBILITY WITH VARIOUS ARDUINO BOARDS

There are numerous Arduino board form factors, and only some of them support shields. The Arduino Uno, Mega, and Leonardo are among the primary shield-compatible boards; special-purpose boards like the Nano and Fio lack the proper rows of connectors top and bottom.

But even among Arduino boards that are receptive to shields, variations in engineering can make certain shields incompatible. This is the case whenever a shield ties into hardware-specific pins on the Arduino—SPI is the main offender here. In the Arduino, SPI is supported by hardware internal to the Atmel microcontroller chip at the heart of the board. This hardware is mapped to specific pins, and those pins may not be the same between Arduino boards.

On the Arduino Uno, SPI pins are 10 through 13, but on the Mega, SPI pins are 50 through 53. So not only are the pins different, but on the Mega no standard shield connects to these pins. On the Leonardo SPI is provided only on the ICSP header, and only some shields provide connection points to this header.

There are other subtle variations in hardware layout between the Arduino boards, and it pays to always verify that your board is compatible with the shield you wish to use. If the documentation for the shield doesn't specify, ask the maker or seller of the shield before you commit your hard-earned cash to something that may not work.

Starting with the R3 version of the Uno, the Arduino has four additional I/O pins to the left side of the top and bottom row (see Figure 8-4). Many shields do not use these pins, or even have connections to them. However, you'll encounter problems if your shield depends on electrically attaching to one or more of these pins and your Arduino is an older model that lacks the I/O connectors.

Figure 8-4 Additional connection pins are provided on updated versions of the Arduino Uno, starting with revision R3.

Tools for Electronic Construction

You need relatively few tools to build the electronic centerpieces of your robots. To get started you should have:

- A *multimeter*, also called *volt-ohm meter*, *VOM*, or *multitester*. It's used among other things to test voltage levels and the resistance of circuits. Multimeters are moderately priced, and are the basic prerequisite for working with electronic circuits. You can get a basic model for under $10. See the ARB Support Site (Appendix A) for additional information on how to select and use a multimeter with your robotic projects.
- Most tasks beyond simple solderless breadboard wiring require a soldering tool of some kind, such as a *soldering pencil*. Even if you never plan to make your own circuit boards for your robots you still need a soldering pencil for basic electronic chores, such as connecting wires to motors.
- *Handy hand tools* for electronic construction include flush wire cutters, used for cutting off wire flush with the surface of a circuit board; wire strippers for 18- to 26-gauge wire; solder clamp or vise, which serves as a "third hand" holding together pieces to be soldered; a set of flat #00 and #0 bladed and Phillips screwdrivers; and small needle-nose pliers.

 New to soldering? There are numerous online guides to help you learn the art and science of soldering. For a list of my favorites, reference the ARB Support Site in Appendix A.

Understanding Wires and Wiring

Almost every electronic circuit uses wire. There are three primary considerations when selecting the right wire for your project:

Wire insulation Most of the wire used in robot electronics is insulated with a plastic covering. This keeps one wire from touching another, and causing a short circuit. An important aspect of insulation is its color. Use different color wiring to denote what it's used for in your circuit. For example, red wire is often used for the + (positive) battery connection; black wire for the − (negative) connection.

Gauge The thickness, or gauge, of the wire determines its current-carrying capabilities. Generally, the larger the wire, the more current it can pass.

Conductor type Wire is made of one or more strands of metal, and wire is either single or multi-strand. Single-strand wire has just one metal conductor. Multiple strand (or *stranded*) wire has many conductors.

Connecting Things Together

You've read that expanding the Arduino by way of shields requires no special wiring because electrical connection is already made by the pins between the Arduino and shield. But what about circuits, sensors, and other parts that aren't on shields? Connecting these means stretching wires between the subsystems of your robot.

The most common requirement is attaching one or more of the Arduino's I/O pins to a solderless breadboard, circuit board, switch, or other device located elsewhere on your robot, as shown in Figure 8-5.

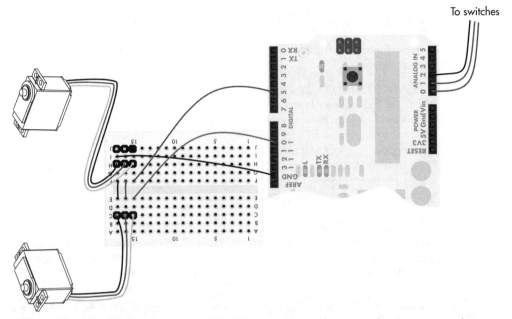

Figure 8-5 The solderless breadboard allows for a way to bridge external components, such as servos and switches, to the Arduino I/O pins.

Wiring is not difficult, and you have several options available, with various costs and levels of complexity between them. The most common approaches are shown in the following table, in approximate order of cost and complexity. There is no single "best" wiring technique, as each has its place.

Technique	Description	Pros	Cons
Soldered point to point	Stretch wires between components, and solder them into place.	Least expensive method as it requires no special connectors; easily cut wires to length to minimize excess.	Difficult to share subsystems between robots; permanent wiring can be cumbersome.
Self-made wire jumpers	Wire cut to length, with attached male or female connector ends. May be constructed using solid or stranded wire.	Wires securely attach between components; wire lengths can be easily customized.	Soldering and/or crimping required; solid conductor wires can break with too much flexing.
Premade wire jumpers	Same as above, but you purchase the jumpers already made.	Same as above, but jumpers come in only certain lengths.	Limited wire length choices, excess must be coiled or wrapped; wire thickness limited to 22 or 24 gauge if inserting directly into solderless breadboard.
Screw terminals	Wires fastened to components using screw terminals. Most commonly used with heavier gauge applications, such as motors.	Low-cost and convenient for higher current jobs requiring larger gauge wires (12 to 16).	Terminals are bulky, and best used for high-current applications.

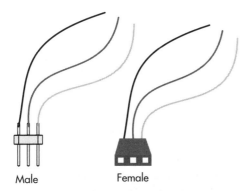

Figure 8-6 Construct male and female header connectors to make it easier to attach robot parts to its Arduino or solderless breadboard.

Male   Female

Much of the wiring on any robot is for carrying low-current signals—analog or digital. These signals can use standard 22-gauge solid or stranded wires. When connecting the robot's subparts, I prefer to use only stranded wiring, as it's more flexible, and therefore not as prone to breaking over time.

Just about *every* robot you'll ever make will require at least one *wire set*. Wire sets are groups of two or more wires that connect the pieces together. Each set consists of the wires themselves, plus some form of termination on the ends.

The most common, and cheapest, termination is the male or female pin. You use a wire with a male pin to connect into any of the Arduino's I/O ports, for example. Wires with male pins also plug directly into solderless breadboards. Likewise, wires with female pins are designed to work with circuit boards with male headers soldered onto them; see Figure 8-6 for examples.

You can purchase ready-made wire sets of a specific wire length and number of connectors. But these tend to be expensive, and the choice of length and number of wires is limited. It's much easier and cheaper to just make your own. There are two popular ways of creating your own wire sets: soldering directly to male header pins, and crimping onto male or female pins. Both techniques are covered next.

CONSTRUCTING WIRE SETS BY SOLDERING TO MALE HEADER PINS

There are many ways to produce your own wire sets. One of the most economical is by soldering the wires directly to male header pins. These pins in turn can plug directly into the Arduino or solderless breadboard. The technique requires moderate-to-better soldering skills, but is relatively inexpensive, as all you need is the wire plus a length of double-length (same length both sides) of male header pins that you snap to the desired length. Figure 8-7 shows the general concept.

1. Start first by cutting some 22- or 24-gauge insulated stranded conductor wire to the desired length. Don't be stingy with the wire, but don't make it so long the extra gets in the way. Give yourself an additional inch or two so you can twist the leads together to make a nice pigtail.
2. Strip approximately 1/8" to 3/16" of insulation from both ends of the wire, and use your soldering pencil to pre-tin the wire (tinning means to apply a very small amount of solder).
3. Leave the header pins in the long strip for now, and clamp in a soldering vise or "third hand" soldering aid. Locate the pins you want to solder to, and pre-tin one end of each pin—don't tin the other end as that'll go into the connector.
4. After tinning is complete carefully tack-solder the wires to the pins. Do just one end of the wire set at this time.

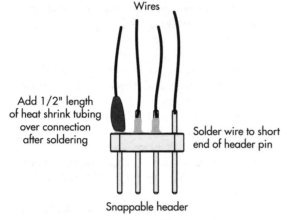

Wires

Add 1/2" length of heat shrink tubing over connection after soldering

Solder wire to short end of header pin

Snappable header

Figure 8-7 The concept of soldering wires to a male header pin. First tin the wire and pin ends with solder, then lightly tack the wire to the end of the pin. Finish by applying a small piece of heat-shrinkable tubing.

5. Apply heat-shrink tubing to finish off the soldered ends. The tubing makes for a more professional look, plus it helps prevent short circuits. Cut lengths of tubing to about 1/2", insert over the wires and pin ends, then use a small heat gun to shrink. (In a pinch you can use the flame of a small butane lighter, but exercise care not to melt the tubing, or catch anything on fire!)

6. Repeat the previous steps for the pins on the other end of the wire, except thread the heat-shrink tubing over the wire first. When soldering is complete, position the tubing into place and apply by shrinking with heat.

7. Once soldering is complete snap off the pins you want to use.

CONSTRUCTING WIRE SETS BY CRIMPING TO HEADER PINS

Another (and frankly easier!) way to make your own wire sets is to use special crimp-on male or female pins. You'll need a special crimping tool to make this work. Once the header pin has been attached to the wire, you then insert it into a plastic *shell*. This shell comes in various sizes, designated by the number of pins in each row and by the number of rows.

For example, 1×1 means a single pin in the shell; 1×2 means a single row of two pins; 2×3 means two rows of three pins each, and so on. The most common shells are the single-row variety, such as 1×3, used to construct servo extension wires. Buy just the shells you use most often. You can always use multiple shells to create sets with any number of wires—for instance, a 1×2 and a 1×3 to make a combined connector with one row by five pins.

Making a custom crimp-on wiring set is easy. You need to use 22- or 24-gauge wire. Stranded conductor is best.

1. Cut the wire to length, and remove about 1/4" of insulation from both ends.

2. Place the male or female crimp-on pin into the crimping tool, being careful to properly align the pin into the jaws of the tool.

3. Insert the wire into the back end of the pin, then crimp. Test the integrity of the connection by lightly pulling the wire from the pin. If the wire comes out you'll need to remake the connection.

4. Insert the pin into the desired receptacle in the shell. The pin will lock into position when it's pushed all the way in.
5. Repeat steps 1 through 4 for all wires in the wiring set.

You can bypass the crimp-on step by purchasing wiring with the male or female pins already attached. This is useful if you don't want to invest in the crimping tool—a good one costs $30 to $40—or if you simply want the convenience of already crimped wires. You just snap the pin ends into the shell to make your wiring sets. The wiring comes in standard lengths of 3", 6", and 12". Longer lengths are also available. Pololu (see Appendix B) is a primary source of pre-crimped wire.

Truth be told I prefer using the pre-crimped wires, as I can work much faster. Yes, they're more expensive, and I must choose from standard lengths, but the ease of use outweighs both of these limitations.

Making Your Own Circuits

Over the last several years the advent of the "breakout board" has greatly changed the way robots are constructed. Rather than building every circuit from scratch, these boards have all the necessary components already on them; all you do is stretch wires between the board and your Arduino. Example breakout boards include accelerometers, electronic compasses, ultrasonic and infrared sensors, GPS units, and many more.

But sometimes there isn't a breakout board for all the robotic subsystems you want to build. In these instances you must create your own circuit. Techniques include plugging the components into a solderless breadboard—this is the method employed in several projects throughout this book—or you can solder things into pre-drilled boards, and even make your own custom printed circuit boards.

USING A SOLDERLESS BREADBOARD

By now you already know what a solderless breadboard is and how to use it. Suffice it to say, it's the easiest method of building your own circuits for your robot. On the downside, because components are soldered into place, they can eventually work themselves loose. For the average desktop bot, and assuming a well-made and relatively new breadboard, this actually happens less often than you think, but it's still something to be watchful of.

As needed you can keep things in place by tying down the larger and heavier components with small cable ties and rubber bands. Some folks report good results by wrapping food cling wrap—such as Saran Wrap—around the breadboard and components, but I think this could cause more trouble than it's worth. The reason: cling wrap generates lots of static. It also acts as an insulator; the covering may not allow those components that run hot access to room air for necessary cooling.

USING SOLDER BOARD

A solder board comes to you with holes already drilled into it. Many have copper traces on the underside that connect sequences of these holes together. You insert and solder components into the holes depending on how you want to wire your circuit.

Plated—no strips   Strips of 3   Continuous strips

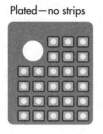  

Figure 8-8   Three common variations of plated solderable printed circuit board

One popular type of solder board is modeled after the layout of the solderless breadboard: holes are lined up in the familiar column and row layout. The columns are interconnected, and you bridge the components together in the same way as you do with a solderless breadboard.

Another type is the stripboard. Several variations are shown in Figure 8-8. Traces join two or more holes together; a common arrangement is one trace segment for a row of every three holes. This allows you to insert the lead of an electronic component, like a resistor or capacitor, and connect up to two wires or other components to it.

Select the style of stripboard that best suits your project and tastes. Variations include:

Plated holes without strips   These are little more than copper plating around a series of holes. You complete the board by attaching wires to the component leads.

Continuous strips that run the length of the board   The strips run in one direction. To make a circuit you solder in components and wires, then use a sharp knife (or a specialty tool) to break the strip after all the connections are made for that particular part of the circuit.

Strips in groups of three to five holes   Each segment spans from three to five holes (sometimes more). Some boards have additional strips running perpendicular to the segments; these are common *busses* for easy connection to positive and negative terminals from the power supply.

MAKING A CUSTOM PRINTED CIRCUIT BOARD

The Arduino itself is constructed on a printed circuit board (PCB). Though far from a requirement, you can make your own PCB for your circuits. The two main methods are creating the board yourself or sending it out to a service bureau.

- To create the board at home you first apply a *resist* to a board made of fiberglass and coated on one or both sides with a thin layer of copper. The resist forms the layout of the circuit. You then immerse the board into a caustic fluid that etches away the exposed copper. Once all the copper has been eaten away, you remove the resist. From there you drill holes for components to poke through.
- To send the PCB to a service bureau you use software to design the layout of the board— you can use a commercial PCB layout program, or many services offer free software. Once the layout is complete, you submit the pattern files that describe your board to the service bureau, who then in turn produces the physical board using automated equipment.

Of the two methods, homebrew boards are the least expensive, but the process is time consuming, and involves working with poisonous materials. PCBs made by a service bureau cost between $5 and $50 each, depending on size, the number of individual boards you order, and other factors. However, the results are top-notch, equaling and even rivaling the finished breakout boards you purchase for your robotic endeavors.

Apart from waiting for it to arrive in the mail, the hardest part about creating your own custom PCBs using a service bureau is learning how to use circuit layout software. These programs are collectively referred to as *eCAD,* which stands for electronic computer-aided design. Some high-end eCAD programs cost thousands of dollars, but you'll probably want to opt for the free or low-cost versions of popular eCAD programs like Eagle and DipTrace. The less expensive versions of these and similar programs differ from their more expensive cousins in the complexity of the PCBs they can build.

Proprietary design software provided by the service bureau has the added benefit of guiding you through the fabrication process particular to that service. That can help minimize costly errors. They can also provide instant cost estimates, based on board size, number of holes that must be drilled, and whether the board is single or double sided.

Arduino users can also take advantage of Fritzing (available at *www.fritzing.org*), a free eCAD program designed for creating printed circuit board shields. By connecting wires on a virtual solderless breadboard in the software, your circuit designs are automatically converted to schematics *and* printed circuit boards in the standard shield layout.

Figure 8-9 shows one of the example projects that comes with the Fritzing software. Simply by clicking tab buttons inside the Fritzing program you can view the circuit as a schematic (Figure 8-10) and as an Arduino shield PCB (Figure 8-11). In all views you can drag components using the mouse to reposition them.

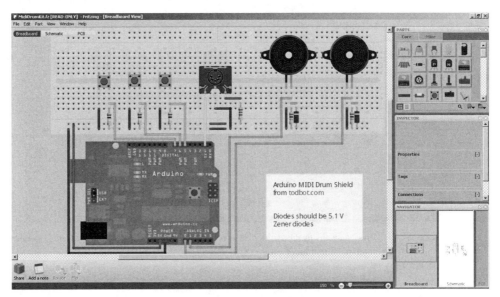

Figure 8-9   One of the example projects that comes with the open-source Fritzing design software, showing a breadboard view of various components connected to an Arduino. This particular project is the MIDI drum kit.

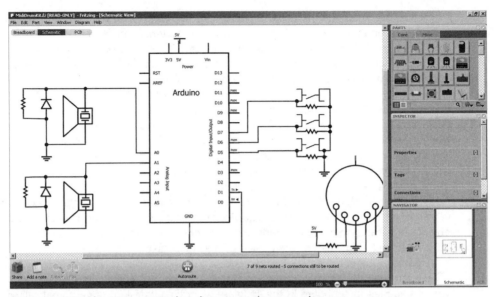

Figure 8-10   The Fritzing MIDI drum kit project, shown in schematic view

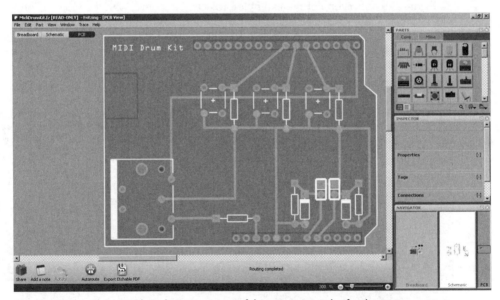

Figure 8-11   The MIDI drum kit project, one of the many examples for the Fritzing open-source design software, shown in printed circuit board view

With the shield layout generated by Fritzing, what's next? You have several options:

- *Export the PCB layout as an etchable PDF.* Use a laser printer to print the PDF file onto a toner transfer film. With a clothes iron you then transfer the toner from the film to the bare copper surface of a circuit board. The board can then be etched as discussed previously.

- *Export the design to an Eagle-compatible file* for use with the CadSoft Eagle PCB software, mentioned earlier. You can then open the file in Eagle for further work.
- *Export the layout to a series of Gerber files* that contain all the information that an automated PCB service bureau can use to prepare your board. The files generated by Fritzing specify the printed circuit layout, solder mask (a coating on the unsolderable portions), silkscreen (text and graphics), and hole sizes for components.

What You Need to Know About Interfacing

Interfacing is the term used to describe connecting external things to your Arduino, such as motors, LEDs, speakers, and even other Arduinos. Sometimes an interface is just a wire; other times it's a complete circuit designed to convert the voltage and current levels provided on the Arduino with those needed to operate some external piece of hardware. A motor controller is a good example of an interface circuit that is used to connect the Arduino to a motor.

While there are literally tens of thousands of doo-dads you can attach to your Arduino, most robotic tasks boil down to a small subset of output and input tasks. Here are the ones you're most likely to encounter.

INTERFACING TO LEDs

You learned how to use the Arduino's onboard light-emitting diode in Chapter 5, "Programming the Teachbot: Seeing It React." This LED is permanently connected to digital pin D13 of the Arduino, and serves as a quick and easy method of providing visual feedback in your sketches.

But what if you want to have more LEDs? It would be easy—but wrong— to just connect the LEDs directly to the Arduino's I/O pins. If you do that, the LED will burn up within seconds. Why? Light-emitting diodes try to draw in as much current as they can, yet they're susceptible to damage if too much current flows through them.

To avoid wrecking your light-emitting diodes you need to insert a resistor inline (called *in series*) with the LED, like that shown in Figure 8-12. The circuit uses a 330 Ω resistor to limit the current.

Selecting the value of the resistor requires some math, plus knowing some things about the typical LED:

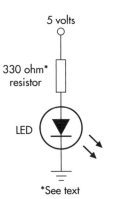

5 volts

330 ohm* resistor

LED

*See text

Figure 8-12 Proper connection of an LED uses a current-limiting resistor.

- Most LEDs will burn out if they consume more than about 40 milliamps (40 mA). We want to ensure the LED gets less than this amount of current. For the purposes of this example, we want to have the LED receive no more than 10 mA.
- You also need to know the *forward voltage drop* across the LED. This is literally the amount of voltage that is lost when current is passed through the component. The typical voltage drop of an LED is 1.5 to 2.5 volts, though this varies depending on LED type and color. The typical voltage drop for the typical standard-brightness red-colored LED is around 1.75 volts. This is the figure we'll use in calculating the desired resistor value.

Apply this simply formula to determine the value of the resistor:

```
(Vin - Vdrop) / mA = R
```

Vin is 5, and Vdrop (the voltage drop) is 1.75. We want to limit current to 10 milliamps, so the formula becomes

```
(5 - 1.75) / .010  = 325
```

or

```
3.25 / .010 = 325
```

Notice the current draw for the LED is a decimal fraction: the formula assumes amps rather than milliamps. There are 1000 milliamps in one amp.

Resistors with four color bands come in only specific standard values, so you need to find the nearest one to your target of 325 ohms. Picking the next higher standard value is safer, since the worst thing that'll happen is that the LED won't glow quite as brightly. The next higher standard value to 325 is 330 Ω.

The current flowing through an LED regulates its brightness. A lower current makes the LED glow more faintly, but it conserves power. The less current that goes to each LED in your project, the longer the batteries will last. Of course, there are times when you want very bright lights on your bots. In this case, pick an ultra-bright LED, and run it at a higher—but still safe—current.

Each of the Arduino's I/O pins can safely deliver a maximum of 40 mA, so be sure never to exceed this amount for any LED on your robot. In addition, the entire current demand through the ATmega microcontroller on the Arduino is limited to more than 200 mA, which equates to just five LEDs driven at 40 mA each.

If you need to drive lots of light-emitting diodes, pick a current-limiting resistor for a much lower current, or use a separate driver circuit, like a transistor. See the ARB Support Site (Appendix A) for a discussion on building simple driver circuits you can use when you need to provide a tad more current for your LEDs and other components.

INTERFACING TO MOTORS

As most robots use at least one motor, interfacing the Arduino to motors is a common requirement. How the motor is interfaced depends on the type of motor. From an electrical standpoint, motors vary in two important ways:

Voltage rating. This is the voltage (expressed in volts) specified for the motor. Most motors can be operated at above or below its rated voltage. At higher voltages the motor becomes more powerful but it may heat up more quickly. Avoid operating the motor at greater or less than 25 percent of its rated voltage.

Current consumption. This is the amount of current, in amps or milliamps, drawn by the motor. Higher currents require bigger batteries or power supplies. The current consumption of a motor varies depending on load: the current is lowest when the motor is free-running (*no load*). Current increases as the motor is loaded with more work to do. Current is at its highest, called *stall current,* when the motor is energized but prevented from turning.

Nearly all motors require too much current for the Arduino to handle by itself. Trying to directly connect a motor to the Arduino will result in the motor simply not working, and may cause permanent damage to your Arduino. How the motor is interfaced to the Arduino depends on what type it is.

Interfacing to an R/C Servo Motor

R/C servo motors are the easiest to interface to the Arduino because a driving circuit is built into them. Figure 8-13 shows both a minimum hookup diagram and an optional connection using a 470 Ω resistor. The resistor serves to protect the Arduino in case something catastrophic happens inside the servo and it short-circuits. The resistor drops the maximum amount of current that can be drawn from the Arduino.

Note that some servos may not work, or may not work reliably, with the resistor in place. If this happens, try a resistor with a slightly lower value.

It's considered rare for a servo to become so deranged that it will pose a danger to your microcontroller. Most well-made servos already have these protections built in, but you can add it if you'd like the extra insurance.

The series resistor between Arduino and servo can also serve as a form of noise reducer. In some cases, electrical noise from the servo can literally travel back to the Arduino through the signal wire, causing a disruption inside the Arduino. I've seen this, for example, when using piezo touch sensors with servos. Piezo devices have a high impedance, and when coupled with the Arduino's ADC pins, the noise from the servos can cause cyclic (happens about once a second or two) interference.

Apart from adding a buffer circuit, like an op amp, between piezo device and the Arduino, the effect can be minimized by inserting a resistor of about 1K in line with the signal wire to each servo. You may also wish to add a 47 µF or higher electrolytic capacitor across the servo power pins.

Remember that you can't just apply + and – voltage to an R/C servo and expect it to run. The servo must be supplied with precise pulses. This is covered in Chapter 4, "Programming the Teachbot: Making It Move."

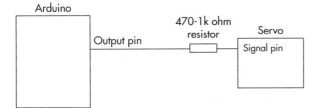

Figure 8-13 A resistor may be used on an Arduino output pin to help prevent possible damage to the pin should the external device try to draw too much current.

Interfacing to a DC Motor

A motor that operates from *direct current*—like the kind from batteries—is called a *DC motor*. They're the most common kind of motor in robotics; in fact, inside that servo motor discussed earlier is really a DC motor.

In most cases the DC motors you use will be attached to a gearbox, which reduces the speed of the motor to something more usable for your robot. This gearing down also makes the output of the motor much more powerful.

Suffice it to say you can't connect one directly to an Arduino, as the Arduino cannot supply enough current to properly operate the motor. There are three primary methods for interfacing a DC motor to the Arduino: relay, transistor or driver, or H-bridge.

- A *relay* is an electrically operated switch. The Arduino controls the relay, which in turn is connected to the motor. Despite their age and electromechanical construction, relays are still useful in robotics, and they're easy to use. Clever use of relays allows you to turn motors on and off, and change their direction.
- A *transistor* can be used as an off/on switch. Like the relay, the Arduino operates the transistor, which in turn controls the motor. A single transistor can only turn a motor on and off. To reverse the direction of the motor you need at least two transistors, though the most common approach uses four. See H-bridge, next. (Transistor-driving circuits may also be used for many other forms of output interfacing, such as connecting the Arduino to a large speaker or operating a solenoid.)
- An *H-bridge* is a combination of four transistors—the name comes from the H shape of the connections when the transistors are drawn on paper. H-bridges are all-electronic devices that can control the operation and direction of motors. You can build an H-bridge using separate transistors, but a better and easier approach is to use a self-contained H-bridge module. The design of H-bridges is complex, and modules greatly simplify the electronics, while avoiding common pitfalls of homebrew circuits.

See Figure 8-14 for an example of connecting the one motor to the popular L293D H-bridge module. Shown is only one motor attached; the L293D supports two motors. To run the motor and change directions you apply LOW or HIGH signals from the Arduino to the Input 1 and Input 2 signals of the L293 H-bridge module.

See Chapter 9, "Enhancing the Teachbot," for a hands-on guide on interfacing an Arduino to a motor control circuit. Also visit the ARB Support Site (Appendix A) for additional online resources available to you.

INPUT PROTECTION

LEDs and motors are examples of output interfacing. There's input interfacing, too. Interface techniques for inputs take into consideration voltage and current levels exceeding what the Arduino is designed to accommodate. The Arduino has basic input protection on all I/O pins to guard against applying voltage below ground (0V) or above V+ (5V for most Arduino

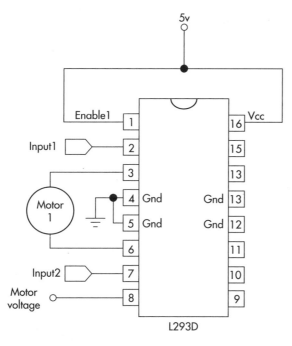

Figure 8-14 Sample wiring diagram for the popular L293D motor bridge integrated circuit. This example shows the wiring for just one motor; the L293D can independently operate up to two motors.

boards; or 3.3V for those that work at this lower voltage). However, in some cases additional simple circuitry is advisable to provide an added margin of protection.

The simple circuits shown here require only minimal components. Most don't change the behavior of the Arduino.

There's no need to bother with adding more input protection for simple interfacing chores, like adding switches for touch sensors, or phototransistors for detecting light, as they are not likely to introduce voltages or currents beyond a safe level.

The input protection methods that follow are worthwhile when using homebrew circuits that may malfunction, or when interfacing unusual gadgets to the Arduino that weren't meant for connection to delicate electronics—like the output of a smoke alarm. The more you experiment with strange sensors, the more you'll want to use protection.

Current-Limiting Resistor

A resistor inline (in series) with the I/O pin will limit current from an external source from getting into the Arduino, as shown in Figure 8-15a. The value of the resistor depends entirely on the application, but can be as low as 22 Ω and as high as 10 kΩ. You might start with a higher value. A 470 Ω resistor is shown for example purposes only.

External Clamping Diodes

Two low-cost signal diodes may be placed on the input pin as shown in Figure 8-15b. Most any signal diode will work; the 1N5817 Schottky fast-acting diode is a popular choice. It's good for up to 20 volts at one amp.

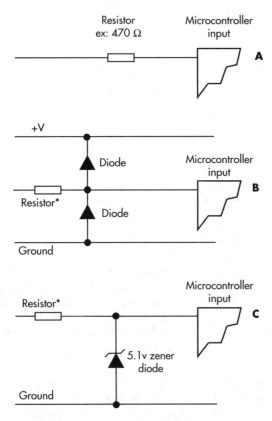

*See text; exact value of resistor depends on application

Figure 8-15 Three common methods for providing minimal input protection between external devices and the Arduino

External Voltage-Limiting Zener Diode(s)

An alternative to using two signal diodes is one zener diode, as depicted in Figure 8-15c. Zener diodes are unusual in that they start conducting at a specific voltage. When used as input protection, select a voltage rating for the zener that's close to the operating voltage of the Arduino, which is either 3.3V or 5V, depending on the model of board you have.

Here are some parts numbers of suitable zener diodes. These are rated to 500mW, which should be sufficient for most protection chores.

Zener	Voltage
1N5226B	3.3V
1N5231B	5.1V

How do you know what value of resistor to use? It all depends on the current passing from the source to the input. For most signal input protection tasks, the current is usually quite low, so the value of the resistor can be fairly low, in the neighborhood of 30 Ω to 1K. It's beyond the scope of this chapter to detail the math formulas to use to arrive at the optimum value of the current-limiting resistor. Check out Appendix A for the ARB Support Site for additional resources on this subject.

Typical Applications for Input Protection

If all this added input protection isn't (usually) needed for routine interfacing chores, then what *is* it good for? Here are three applications you should consider:

Piezo disc "knock" sensor   Piezo discs are both producers and consumers of electricity. When voltage is applied to a piezo disc, it acts as a sound element. When pressed or banged on, the disc produces a voltage—the level of the voltage depends on how hard you hit the thing. It's quite possible to bang on a piezo disc hard enough that it produces hundreds and even thousands of volts.

   Normally, this would obliterate your Arduino, except that the current accompanying this voltage is very low. The Arduino's internal input protection usually handles it without complaint. But in some cases (like banging the disc with a hammer), the voltage can be high enough to produce a spike that could damage your Arduino. Figure 8-16 shows a knock sensor protector, plus a 1M Ω resistor wired in parallel with the disc to make the circuit more sensitive.

Smoke alarm sounder interface   The innards of battery-operated smoke alarms can be hacked to create robot smoke detectors. Disconnect the wires going to the piezo speaker, and reconnect them to an Arduino input. Use the two diode or single zener circuit, shown earlier, to limit the voltage and current that can be introduced into the microcontroller.

Bare-wire whiskers   You can make your own whisker-style touch sensors using springy wire. When the wire touches something it bends slightly, making contact with an electrical post. The wire and the post complete a circuit; this circuit is connected to one of the Arduino's inputs.

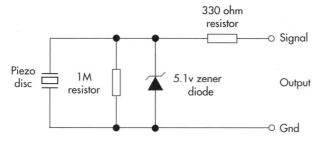

Figure 8-16   Suggested wiring arrangement for providing input protection to the Arduino when using piezo discs. The resistor and zener diode help to limit both current and voltage, and are particularly handy when the piezo disc is subject to strong impacts.

On dry days the long bare wire can actively collect static electricity, acting something like a static antenna. Since the wire is directly connected to the Arduino, it's possible for this static to be conducted right into the Arduino. Again, the Arduino's internal circuitry guards against most problems, but the addition of an external resistor to limit current, plus clamping diodes to limit voltage, affords even better protection.

Reducing Electrical Interference for Inputs and Outputs

Motors and other electronic components can generate electrical interface that may affect the operation of your Arduino. This noise often travels from or to the Arduino via its I/O and power pins. You can reduce or eliminate most noise by following some simple steps.

WAYS TO REDUCE NOISE FROM MOTORS

Electric motors can generate copious amounts of noise, both via the wires that connect things to your bot, and literally through the air. You can also use small ceramic capacitors to filter out much of this noise. The capacitor literally "soaks" up the electrical transients produced by the motors, which in turn reduces the effect of the noise on other parts of the circuit.

- Place a 0.1 µF ceramic capacitor directly across the motor terminals. Just solder the capacitor between the terminals, along with the power leads to the motor.
- If the motor still generates too much electrical noise for the circuit, then as an extra precaution solder a 0.1 µF ceramic capacitor from each power terminal to the metal case of the motor.

 Soldering the capacitor to the case can be a little tricky. To get the solder to stick to the motor case, be sure to clean the case and rough up the metal with a small file. Dab on some solder flux paste over the area, and turn up the heat of the soldering tool.

WAYS TO REDUCE NOISE FROM COMPONENTS

All electronic components generate their own noise, including the Arduino itself. Electrical noise is typically passed through the circuit via the power supply connections. For this reason, it's customary to add so-called bypass or decoupling capacitors near circuit components that either produce electrical noise, or may be adversely affected by it. (By the way, the terms bypass and decoupling refer to the role the capacitors play, not to a specific type of capacitor. Numerous forms of capacitors will work here, as noted next.)

A ceramic or tantalum capacitor is often used for decoupling immediately beside ICs and other noise-producing components. To do the most good, the cap is placed as close to the power pins of the component as possible. For most noise-absorbing work, pick a capacitor with a value of 0.01 µF to 0.1 µF (10 nF to 100 nF).

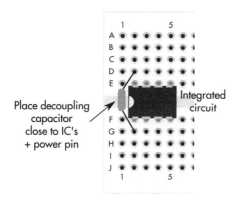

Place decoupling
capacitor
close to IC's
+ power pin

Figure 8-17   Recommended placement
of decoupling capacitors near the power
supply pins of an integrated circuit

If your project uses several individual circuit boards, you may wish to also use a 1 μF to 10 μF polarized tantalum capacitor, stretched between the V+ and ground connections as they enter the board. Repeat this for every circuit board you use. Figure 8-17 shows how these capacitors are placed.

The Arduino already has decoupling capacitors on it for its power pins, so you don't need to add them separately. But even though the Arduino contains its own decoupling capacitors, it's always a good idea to add them to the circuit boards you make for your projects.

Most commercial shields and accessories for the Arduino have decoupling capacitors, but if a board you're using doesn't, and it seems electrical noise is causing problems, you may wish to add one or more capacitors across the power supply connections to see if that helps.

Part 3

Hands-on Arduino Robot Projects

Chapter 9

Enhancing the Teachbot

Radio control servos make great robot motors. But they suffer from two disadvantages: as robot drive motors they tend to be a little slow, and control over the full range of speed is limited.

Enter the DC gear motor, available in literally thousands of sizes, styles, and other permutations. With the right gear ratio, DC gear motors can zip your robot across the floor; with the right electronics, DC gear motors provide a high level of control of that speed.

The Teachbot DC, described in this chapter, demonstrates the fundamentals of using DC gear motors to drive a robot. It's designed around the same base as the Teachbot Servo described in Part 1 of this book. But instead of being powered by R/C servo motors, this new version uses two relatively inexpensive miniature gear motors to do the steering and locomoting.

Electrical power for the robot is provided by a 9.6 volt rechargeable battery pack, the kind you'd use in an R/C race car. An Arduino-compatible microcontroller operates the motors via a low-cost dual motor controller board that you buy ready-made. As with the Teachbot Servo, this version doesn't involve the construction of any complex circuitry. Everything is off-the-shelf, readily available, and easily swappable should you wish to try something else.

 Be sure to visit the ARB Support Site (see Appendix A) for additional code examples and variations. All sketches from this book are available for ready download.

General Parts List

The Teachbot DC uses the same bottom deck as the Teachbot Servo. In fact, it uses two of the bottom decks, stacked one on top of the other, separated by 1" plastic standoffs. The basic version of the completed Teachbot DC is shown in Figure 9-1.

As already noted, rather than using R/C servo motors, Teachbot DC relies on a pair of miniature precision DC gear motors. Despite their diminutive size, these motors deliver considerable torque. They can easily accelerate the robot to speeds of several feet per second. Complementing the motors are a pair of two 60mm (about 2-3/8") diameter rubber-treaded wheels, plus two casters outfitted with a 1/2" plastic ball. The wheels are molded to directly fit onto the D-shaped 3mm output shaft of the micro motors.

For your convenience, all the main mechanical parts are available from a single source, as noted in the following parts list table. You can, of course, substitute any or all of these parts for others of your choosing. But if you do, be sure to select components that provide the same general floor clearance—about 1/2" to 3/4"—as the stock Teachbot DC. Optional features of the Teachbot DC, such as a line-following module, require the floor clearance space.

Figure 9-1 The basic Teachbot DC, with solderless breadboard and Boarduino Arduino-compatible microcontroller

The main components of the Teachbot DC are:

2	7" base decks, same as bottom deck of Teachbot Servo (see Chapter 3 for construction details)
2	Micro DC gear motors, 100:1 gear ratio, 3mm diameter D shaft: Pololu 1101
1	Motor mount bracket for micro DC motor (pair): Pololu 989
1	60mm × 10mm wheels with 3mm D shaft axle hub (pair): Pololu 1420
2	1/2" diameter ball casters: Pololu 952
1	Adafruit Boarduino, breadboard-ready Arduino, with USB and ATmega328: Adafruit ID #19
1	Dual serial motor controller, model Qik 2s9v1; 1 amp continuous drive capacity: Pololu 1110
1	9.6V rechargeable (NiCd or NiMH) 6-cell AA battery pack, with disconnect, 4-cell × 2 rows (see later)
1	Double-pole double-throw (DPDT) miniature toggle switch, center-off
1	Solderless breadboard, 840 contact points, +/− power bus rails on both top and bottom
Misc (see text)	5 volt 1 amp LM7805 linear regulator (TO220 package); 220 μF aluminum or tantalum electrolytic capacitor; 0.1 μF ceramic disc capacitor; double-long male pin headers (see text); precut (3", 6", 12") insulated wiring with crimped female connectors; header shell assortment; breadboard jumper wires; construction and assembly hardware as specified in text

The battery is a self-contained pack, the kind designed for high-speed R/C race cars. You can purchase it ready-made, or make one yourself if you're familiar with the technique (most any book or site about model car racing will show you how). The layout of the batteries should be eight AA cells, arranged in two rows of four cells each, as shown in Figure 9-2. Avoid any other arrangement, or else the pack won't fit properly on the robot.

The leads of the pack should be terminated with a polarized quick-disconnect. Mine used a standard-size Tamiya battery quick-disconnect, but you can use any other, as long as you have the mating connector. You need to be able to unhook the battery from the robot so you can replace or recharge it. You can use either NiCd or NiMH rechargeable cells in the pack, but be sure to use a suitable charger for the type of batteries you are using.

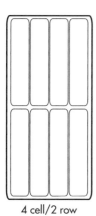

4 cell/2 row

Figure 9-2 The eight-cell battery must be configured as two rows of four cells each.

Constructing the Teachbot DC

Begin by cutting and drilling two Teachbot bottom decks, as described in Chapter 3, "Building the Teachbot Learning Platform." You only need to construct the decks; the servo mounts are not used in the Teachbot DC version. Referring to Figure 9-3, on one of the decks:

Drill holes for the two motor mounts:

1. Mark the position for the holes for each motor mount by first placing the wheels on the motors (to ensure proper spacing between the deck and wheels). Snap the motor mount brackets over the open gear portion of the motors. The "Pololu" name should face toward the terminal end of each motor. (If the mounts are placed the other direction, the gears may jam against the underside of the mount housing.)
2. After marking, drill the holes for the motor mounts using a 1/8" drill bit. Yes, the bit is a little oversized for the #2 screws you'll use to secure the mounts, but this is a good thing, as it allows some extra play for aligning the motors.

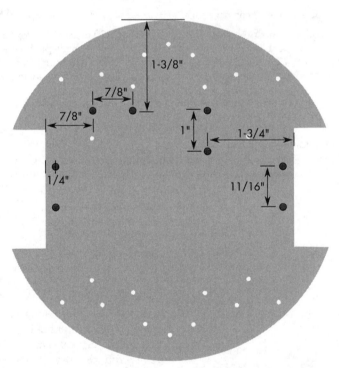

Figure 9-3 The Teachbot DC uses the same layout for its bottom deck as the Teachbot Servo (see Chapter 3). Drill additional holes as shown.

Drill holes for the front and back ball casters:

3. Mark the position for the 1/2" ball casters using the spacers included with each caster. The center of the casters should go through the centerline of the deck. Exact orientation of the holes isn't super critical.

4. After marking, drill the holes for the ball casters using a 1/8" drill bit. The holes will be slightly oversized, which is acceptable.

Drill holes for the serial motor driver and power switch:

5. Mark and drill (using a 1/8" bit) the holes to attach the serial motor driver. Mount the serial driver using two 2-56 × 1/2" machine screws and 2-56 nuts, with two 1/8" spacers—hole size isn't important; #2 through #4 will work.

6. Construct a power switch panel from 1/8" plastic (PVC, ABS, acrylic; see Figure 9-4). The center hole is sized to accommodate a panel mount miniature toggle switch with a 1/4" threaded bushing. Adjust the size of this hole if using a switch with a different diameter bushing. Assemble the metal brackets on the power switch panel using 4-40 × 3/8" machine screws and 4-40 nuts. Don't mount the switch into the panel just yet.

7. With a 1/8" bit, mark and drill two holes in the deck for mounting the switch panel. Attach the power switch panel to the deck using 4-40 × 1/2" flat-head machine screws and 4-40 nuts.

Attach motors and ball casters:

8. Attach the motors to the base using the motor mount brackets and 2-56 × 7/16" machine screws and 2-56 nuts (see Figure 9-5). The head of the screws should come

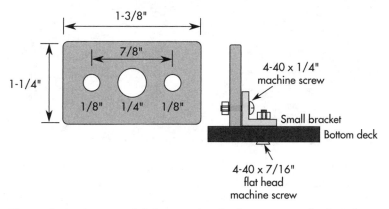

Figure 9-4 Cutting and drilling template for the power switch. The hole in the center is sized for a mini toggle switch with a 1/4" threaded bushing.

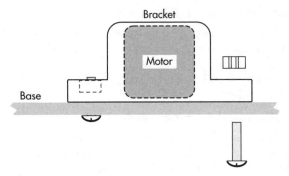

Bracket

Motor

Base

Figure 9-5 Assembly detail for the motors and motor mounts

up from the bottom of the base. Be sure the wheels run parallel to the inside well of the base.

9. Attach the ball casters to the front and back of the base using 2-56 × 7/16" machine screws and 2-56 nuts (see Figure 9-6). The casters come with 1/16" and 1/8" spacers. Use the 1/8" spacer for the front caster (the end that the battery goes over); don't use a spacer for the rear caster. The screw heads should come up through the body of the caster.

 The fastener hardware is included with the mounts and ball casters, but you may wish to get a few extra, as their small size makes them easy to lose.

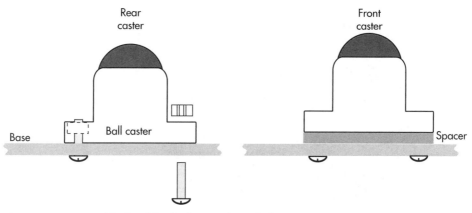

Figure 9-6   Assembly detail for the front and rear ball casters

Mount the battery pack to the bottom deck:

10. Use two or three 1" Velcro squares to attach the 9.6 volt battery pack to the bottom deck. As shown in Figure 9-7, the top of the battery pack should be approximately 1" from the serial motor driver—the Velcro allows you to reposition the battery a bit if you get it too close or too far away from the driver board.

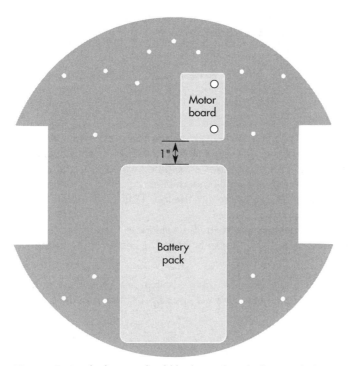

Figure 9-7   The battery should be located on the bottom deck, positioned about 1" from the motor driver board.

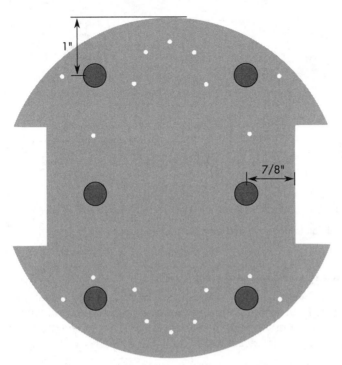

Figure 9-8   Use an additional Teachbot Servo bottom deck (see Chapter 3), with 1/2" holes drilled as shown. This allows wires to pass through from the bottom deck.

Prepare the top deck:

11. In the second deck, drill additional 1/2" diameter holes, as shown in Figure 9-8, for passing through wires between the top and bottom decks. The exact location of these pass-through holes isn't critical, but they should be located more or less as shown so they don't conflict with hardware you'll mount later.
12. Use 4-40 × 1/2" machine screws to secure the four 1" nylon standoffs to the underside of the top deck.
13. Use three 1" Velcro squares to attach the solderless breadboard to the top deck (Figure 9-9). The breadboard will fit evenly down the center of the deck.

 As needed reorient the power buss rails on the top and bottom of the breadboard so that the V+ (red) row is along the top for both rails. This will ensure your breadboard matches that shown in this book.

Figure 9-9   Mount the breadboard to the top deck using 1" squares of Velcro.

Wiring the Teachbot DC

With the hardware of the Teachbot DC finished, you're now ready to wire things up. For my prototype I elected to use precut wiring with crimped female connectors already attached. The connectors then slip into plastic female header *shells,* which are available in various group-ings, such as 1 × 1, 1 × 2, or 1 × 4. The connectors easily snap into place within the shell, forming a solid lock. Although you can use any wiring scheme you like, I recommend this approach, as it's simple, inexpensive, and requires no soldering.

The precut wiring with crimped connectors is available (from Pololu and other sources) in standard lengths. Get a rainbow package of 3", 6", and 12". Also get several sets of double-long snappable headers; these are used as "gender changers" when plugging a female header into a female pin socket, such as the solderless breadboard. See Figure 9-10 for an example.

While you can get precut wiring with male connectors, that starts getting expensive; plus I've found there's a better physical connection when using the female connectors with the snappable male header pins.

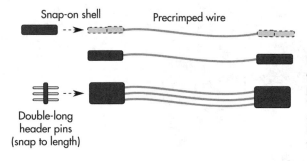

Figure 9-10 Suggested wire and connector scheme for the Teachbot uses precut wiring already crimped with female connectors. Add snap-on header shells to complete. The double-long header pins let you plug the wiring into the breadboard.

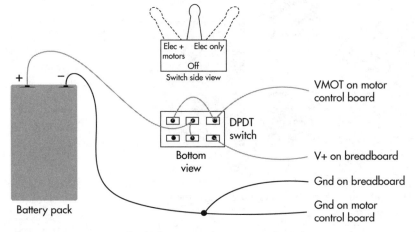

Figure 9-11 Wiring for the battery and power switch. In the center the power is off. Toggling either way powers only the electronics, or both the electronics and the motors.

POWER SWITCH WIRING

Start first with the wiring for the power switch, shown in Figure 9-11. The wired switch is shown in Figure 9-12. After wiring, mount the switch on the bracket you previously constructed.

Main power switch positions are:

- Center: Off
- Right*: Power to electronics only
- Left*: Power to both electronics + motors

BREADBOARD WIRING

See Figure 9-13 for wiring the Boarduino and Qik motor control board. Note the two 9.6V voltage supplies. Both are from the same 9.6V battery; they refer to the two power switch positions, as noted in the section earlier.

---

*Right and left depend on how you wire and orient the switch. It doesn't really matter which direction does what, so long as you're consistent in how the power switches are oriented in your various robot projects.

Figure 9-12 Wired and mounted power switch

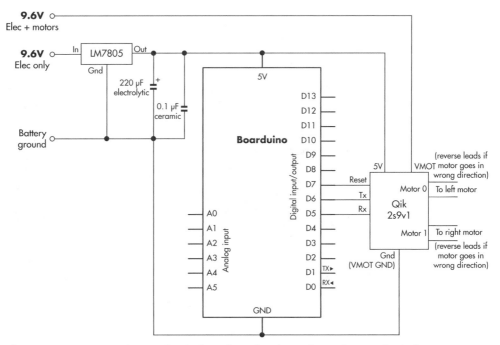

Figure 9-13 Wiring diagram for the breadboard and Boarduino. The Boarduino derives its power from the LM7805 voltage regulator. DO NOT connect the 9.6V battery directly to the Vcc pin of the Boarduino.

The Qik 2s9v1 and similar boards have multiple ground connections, both for logic ground and for motor (VMOT) ground. As the Teachbot DC only uses one battery, you can use just one ground connection, but as noted in the Qik manual, be sure to select the GND pin nearest the VMOT pin. Also be sure not to apply 9.6V to the VCC (logic voltage) input. The VCC pin is for the 5V power provided by the LM785 voltage regulator on the breadboard.

Prior to seating the Boarduino on the breadboard you need to solder pin headers so that the pins face down (they plug into the breadboard). Refer to Figure 9-14 for a component placement guide. There are a couple of additional jumpers that are used with the advanced add-ons described later in the chapter. Add them now so you don't have to pull the Boarduino off to do it later.

FINAL CHECK AND CONNECTION

Double-check your work to ensure against short circuits before connecting the battery. If all looks good, connect the battery, and seat it into position on the bottom base, being careful to tuck any excess wires around the body of the pack. For your reference, the finished bottom base, with battery, is shown in Figure 9-15. Refer back to Figure 9-1 for the finished top base, with Boarduino and wiring installed.

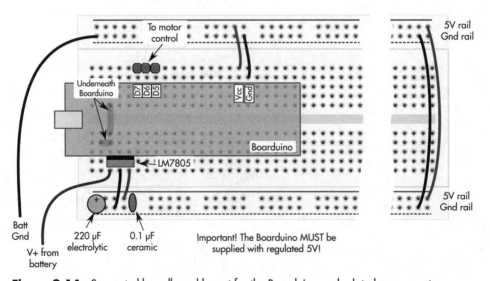

Figure 9-14   Suggested breadboard layout for the Boarduino and related components.

Figure 9-15 Finished bottom deck, showing battery, motors, wheels, motor control board, and switch

Running the Teachbot DC Through Its Paces

With the Teachbot DC assembled and wired, you can run it through its paces to make sure everything is working the way it should. Begin by compiling and uploading the *TeachbotDC_ Basic* sketch to the Arduino.

Leave the main power switch on the Teachbot DC in the center off position. During download, the USB cable from your PC will provide power to the Arduino and other electronics.

As of this writing, the Boarduino (as specified in the "Basic Parts List," earlier) uses the Duemilanove bootloader, so select Arduino Duemilanove Or Nano w/ ATmega328 from the Tools | Board options before compiling.

The TeachbotDC_Basic sketch, and all the sketches for the Teachbot DC, use the *CompactQik2s9v1* library, originally provided by Pololu. Support libraries come and go; if it's no longer available on the Pololu Web site you can obtain it—with the full demonstration code from this book—from the ARB Support Site (see Appendix A). The version of the *CompactQik2s9v1* library on the ARB Support Site has been updated to be compatible with the Arduino 1.0 IDE software.

101010
010101
101010
010101

TeachbotDC_Basic

```
#include <CompactQik2s9v1.h>      // From pololu.com or ARB Support Site
#include <SoftwareSerial.h>


#define rxPin 5
#define txPin 6
#define rstPin 7

SoftwareSerial mySerial =  SoftwareSerial(rxPin, txPin);
CompactQik2s9v1 motor = CompactQik2s9v1(&mySerial,rstPin);

void setup() {
  mySerial.begin(9600);
  motor.begin();
  motor.stopBothMotors();
}

void loop() {
  // Both motors forward
  doDelay();
  motor.motor0Forward(127);     // Left motor
  motor.motor1Forward(127);     // Right motor
  delay (2000);

  // Both motors reverse
  doDelay();
  motor.motor0Reverse(127);
  motor.motor1Reverse(127);
  delay (2000);

  // Turn left
  doDelay();
  motor.motor0Reverse(127);
  motor.motor1Forward(127);
  delay (2000);

  // Turn right
  doDelay();
  motor.motor0Forward(127);
  motor.motor1Reverse(127);
  delay (2000);
}

void doDelay() {    // Short delay between transitions
  motor.motor0Forward(0);
  motor.motor1Forward(0);
  delay (250);
}
```

Once the sketch is downloaded, remove the USB programming cable, and place the Teachbot DC on the ground. Place the main power switch on the robot to the electronics + motors position. After a brief reset period, the robot should go into a short self-test mode demonstrating motor function. The self-test demonstrates the robot moving forward and back, and left and right. All motions are done at full motor speed.

Testing Motor Speed Control

A main benefit of using DC gearmotors to drive your robot is the ability to control their speed over many steps by using *pulse width modulation* (PWM). The Arduino has PWM capability built in—using the *analogWrite* function—and this is the usual method for controlling the speed of motors.

But as we're using a self-contained serial driver (the Pololu Qik 2s9v1) to operate the motors, the PWM speed control is handled by the driver. And the whole process is simplified through the use of a function library provided for the Pololu Qik board. To change the speed of a motor, simply use a lower number for the *speed value* parameter on each motor control statement; e.g., the 127 in the line *motor.motor0Reverse(127)*. Speed is set using a number from 0 to 127:

Speed value	What it does
127	Applies full speed.
0	Stops the motor.
1–126	Varies speed of the motor; the lower the value, the slower the motor.

Depending on the motors used, the weight of the robot, the charge left on the battery, and the friction of the surface the robot is operating on, speed values under 63 may not provide enough torque to move the motors. That means the effective speed curve may not be linear from 1 to 126. Experiment to find the lowest practical speed value that still causes the robot to move.

Try *TeachbotDC_Speed* to experiment with how various settings affect the speed of the robot. In the sketch the speed of both motors are set the same; you can cause the robot to turn in lazy arcs by adjusting the speed of one motor relative to the other. In fact, this very technique is used later in the chapter, in "Adding Wheel Encoders," to automatically adjust the robot's heading to keep it going straight.

TeachbotDC_Speed

```
#include <CompactQik2s9v1.h>     // From pololu.com or ARB Support Site
#include <SoftwareSerial.h>


#define rxPin 5
#define txPin 6
#define rstPin 7

SoftwareSerial mySerial =  SoftwareSerial(rxPin, txPin);
CompactQik2s9v1 motor = CompactQik2s9v1(&mySerial,rstPin);

void setup() {
  mySerial.begin(9600);
  motor.begin();
  motor.stopBothMotors();
}
```

```
void loop() {
  // Both motors forward 3/4 speed
  doDelay();
  motor.motor0Forward(96);
  motor.motor1Forward(96);
  delay (1000);

  // Both motors forward 1/2 speed
  motor.motor0Forward(63);
  motor.motor1Forward(63);
  delay (1000);

  // Both motors reverse 1/2 speed
  doDelay();
  motor.motor0Reverse(63);
  motor.motor1Reverse(63);
  delay (1000);

  // Both motors reverse 3/4 speed
  motor.motor0Reverse(96);
  motor.motor1Reverse(96);
  delay (1000);
}

void doDelay() {    // Short delay between transitions
  motor.motor0Forward(0);
  motor.motor1Forward(0);
  delay (250);
}
```

You may notice that at slower speeds the robot may tend to veer off course a little more often. This is typical, especially when going over carpet or other high-friction surfaces. The reason this occurs is that when going slower the motors don't have as much torque, and not as much ability to resist the friction that might cause resistance to the wheels turning. Later in this chapter you'll learn how to add encoders to the wheels to help the robot go in a straight line.

Adding a Line-Following Module

In Chapter 5, "Programming the Teachbot: Seeing It React," you read about how to add a two-sensor line follower to the Teachbot Servo platform. Two sensors provide the minimum functionality for following lines; more sensors provide more elaborate path-tracking techniques. By adding three or more sensors your robot is better able to discern unusual course changes, even measure the width of the lines.

Adding more sensors is a simple matter of constructing more emitter/detector pairs. Constructing your own sensor pairs is an option, but a better approach—and one that may actually cost less in the long run—is to use a ready-made emitter/detector array. This array, the QTR-8 from Pololu, contains eight sensor pairs, though for the following project we'll only use six of them. Everything comes on the surface mount board, which is thin enough to mount underneath the Teachbot (see Figure 9-16). Just attach power, ground, and signal wires to the sensor pairs you wish to use, as shown in Figure 9-17.

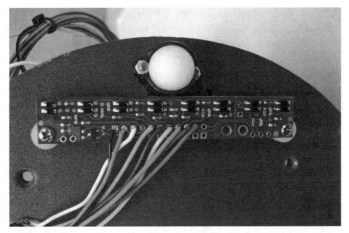

Figure 9-16   Pololu QTR-8 (RC version) IR emitter/detector array, mounted to the underside of the Teachbot. Note the spacers to lower the array closer to the ground.

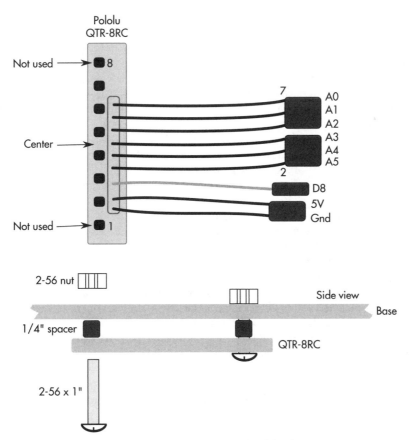

Figure 9-17   Wiring diagram and assembly detail for the QTR-8RC array

Line-following sensors prefer to be as close to the ground as possible. That helps improve accuracy. To get the sensor lower to the ground you'll need to add 1/4" to 3/8" spacers between the underside of the bottom deck and the QTR-8 sensor.

Note that once you add the line-following array the Teachbot will have reduced functionality over carpet. This is because the array may drag across the nap of carpet, impeding motion. If you'd like your Teachbot to be dual use—line follower one day, basic explorer the next— simply remove the array when it's not needed. Be sure the connector shells you use will fit the 1/2" pass-thru holes on the second deck.

Also note that to save I/O pins on the Teachbot, I'm not using all eight sensors on the QTR-8. I've lined up the sensors so that sensors 3 and 4 are centerline over the base. I'm only using the middle six sensors, leaving sensors 1 and 8 unconnected. With this arrangement, I obtain values from sensors 2 through 7.

The QTR-8 also provides an input for controlling its eight infrared emitters. The input allows you to turn the emitters off when they're not needed—that helps conserve battery power. For the Teachbot DC this input line is attached to pin D8, should you wish to use it. Refer to the QTR-8 documentation for code examples.

The QTR-8 is available in two versions: analog and RC. I've selected the RC version, as it can be connected to either the analog or digital pins of the Arduino. The *RC* comes from the resistor and capacitor used with each sensor. In operation, the output of each sensor of the array provides a pulse whose length depends on the relative darkness or brightness of the light detected by the phototransistor. By measuring the length of this pulse, it's possible to obtain very accurate readings from each sensor.

As it turns out, Pololu makes available a handy library for using the QTR-8 array with the Arduino. The library simplifies the use of the sensor array, and it's ideally suited for line following. (Pololu provides several versions of the QTR-8 library at their *Pololu.com* site, including one for generic AVR microcontrollers. Be sure to get the one fine-tuned for the Arduino, and that it is for version 1.0 of the IDE software.)

The Arduino QTR-8 library comes in a zip archive file. Download it to your computer, then:

1. Exit the Arduino IDE if it's already running.
2. Find your Arduino sketches folder, such as My Documents\arduino.
3. Look for and open the *libraries* folder (if there isn't such a folder, create it). Extract the contents of the Arduino QTR-8 library into this folder. This creates a new folder named *PololuQTRSensors*.
4. Depending on the way your zip extraction program works, the files in *PololuQTRSensors* may be located within yet another folder. Move the files out of this inner folder, so they are directly under *PololuQTRSensors*.
5. Start the Arduino IDE. In the IDE, choose **Sketch | Import Library | PololuQTR-Sensors**. (If the *PololuQTRSensors* item is not shown, exit the IDE and double-check that you have unpacked the zip file to the proper place. Restart the IDE and look again.)

When you select the *PololuQTRSensors* library item, the following should be inserted into your sketch:

```
#include <PololuQTRSensors.h>
```

This establishes that you wish to use the *PololuQTRSensors* library, and all its features, in your sketch. *TeachbotDC_LinefollowDemo* shows a demonstration of using the QTR-8RC with the Arduino and Teachbot. It uses the *PololuQTRSensors* library to do most of the heavy lifting. The sketch assumes the QTR-8 has been mounted on what is defined as the "front" of the Teachbot, which is the end with the battery.

The *TeachbotDC_LinefollowDemo* sketch is designed to visually show you how the line-following sensor works. The full example of line following with the sensor with motor control is rather lengthy, and to save page space it's provided for download at the ARB Support Site. See Appendix A for details.

TeachbotDC_LinefollowDemo

```
#include <PololuQTRSensors.h>

#define NUM_SENSORS   6    // Sensors used
#define TIMEOUT       2500 // Wait max of 2500us for sensors
#define EMITTER_PIN   8    // Emitter control on D8

PololuQTRSensorsRC qtrrc((unsigned char[])
  {A0, A1, A2, A3, A4, A5},
  NUM_SENSORS, TIMEOUT, EMITTER_PIN);
unsigned int sensorValues[NUM_SENSORS];

void setup() {
  delay(500);
  int i;
  pinMode(13, OUTPUT);
  digitalWrite(13, HIGH);        // LED to show calibration mode
  for (i = 0; i < 400; i++) {
    qtrrc.calibrate();
  }

  digitalWrite(13, LOW);         // Calibration done; turn off LED
  Serial.begin(9600);            // Set up serial comm
}

void loop() {
  // Read sensors
  unsigned int position = qtrrc.readLine(sensorValues);
  // Sensor values as numbers from 0 to 9:
  // 0 means maximum reflectance, 9 means minimum reflectance,
  //    followed by the line position
  unsigned char i;
  for (i = 0; i < NUM_SENSORS; i++) {
    Serial.print(sensorValues[i] * 10 / 1001);
    Serial.print(' ');
  }

  Serial.print("     ");
  Serial.println(position);
  delay(250);
}
```

To use the sketch, compile and upload it to the Arduino. After upload is complete, display the Serial Monitor window. Watch for the pin 13 LED on the Arduino to light, and when it does, slowly move the front of the Teachbot back and forth across a 3/4" black line. This is the training or calibration phase, which sets up the line-following module to properly read the light levels of your line course.

When the LED goes out, the sensor reads the values from the emitter array, as noted later, and displays them in the Serial Monitor window.

Points of interest in this sketch:

- The sketch creates a single object that represents the QTR-8 sensor array. The name of the object is *qtrrc*. Its *constructor* (the programming statement that creates the object) requires a number of parameters, including the Arduino pins used for connecting to the array and total number of sensors.
- The *PololuQTRSensors* library includes a method for calibrating the sensors. This method is called with the line *qtrrc.calibrate()*. This calibration takes place in the *setup()* function when the Arduino first runs the sketch. During calibration the built-in LED on pin D13 of the Arduino lights up. This is your cue to slowly move the sensor array over the lightest and darkest areas of your line-following course. This stores the minimum and maximum detected levels, and automatically adjusts for differences in the individual sensors. The LED goes out when calibration is done.
- I'm using only the *position* value provided by the *PololuQTRSensors* library, which combines the results from all the sensors into one (you can get the results from the individual sensors, as well). The *position* value indicates the relative position of the array over the line. With six sensors, the value will go from 0 to 5000.

Value	Meaning
0	Line is under sensor 1, or has moved beyond sensor 1
1000	Line is under sensor 2
2000	Line is under sensor 3
2500	Line is straddling sensors 3 and 4 (middle of sensor)
3000	Line is under sensor 4
4000	Line is under sensor 5
5000	Line is under sensor 6, or has moved beyond sensor 6

Incremental values between 0 and 5000 are possible. These indicate the line is between sensors. For example, a value of 2500 means the line is midway between sensors 3 and 4. This location denotes the middle-point of the six sensors, so it's taken to mean the line is directly centered under the robot. Any value less than 2500 means the robot has gone off course to the left, so it should be steered to the right. Conversely, any value more than 2500 means it's gone off course to the right.

Even when the robot is over the line, the *position* value will rarely be exactly 2500. So in the full example I provide on the ARB Support Site I've added a *variance* to the centerline measurement to prevent the robot from constantly "hunting" right and left. With a variance of 300, "centered" means anywhere between 2200 and 2800. You are free to play around with this value. Just change the *variance* variable accordingly.

SLIP TURNS VERSUS SPIN TURNS

The full example (on the ARB Support Site) uses two functions for turning right and left. One set is referred to as *slip,* and the other as *spin.* Here's the difference: In a slip turn, one motor stops while the other continues. In a spin turn, one motor reverses direction while the other continues. As shown in Figure 9-18, the two approaches affect how sharp of a turn the robot can make.

- In a *slip turn,* a differentially steered vehicle like the Teachbot turns in a circle equal to double its wheel base. As the Teachbot has a 7-inch wheel base, that means the robot can trace the path of a 14-inch circle. If you cut out a section of that 14-inch circle to make a curve (and depending on the number and placement of sensors), that curve represents the tightest corner the robot can reliably manage.
- As it happens, curves from a 14-inch circle are still pretty tight. But the Teachbot can manage even better if its wheels rotate in opposite directions in a turn. In a *spin turn,* the vehicle can trace a much sharper path; the minimum curvature will greatly depend on the number and placement of the sensors.

Each type of turn has its place. Slip turns tend to be smoother. Only the tightest corners require spin turns. It's not practical to determine if a corner is too steep when using a two-sensor line follower, but it's quite possible in systems that use four or more sensors. The further away the line is from the center sensor, the sharper the turn.

In a more detailed sketch, you could use the *position* information provided by the QTR-8 (or equivalent) sensor to determine which type of turn is required. For example, if the value is within 1000 points of center, the robot need only correct its course using a slip turn. But if the value is over 1000 points of center, a spin turn will provide a better course correction.

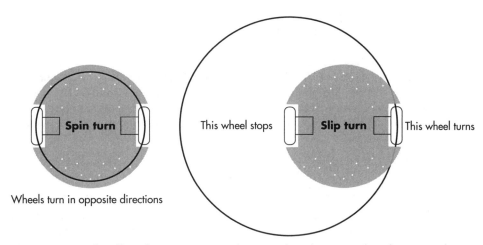

Figure 9-18   The effect of spin turns versus slip turns. The robot can make tighter turns when spinning around its center.

TESTING IF THE SENSOR LEDS ARE LIT

The QTR-8RC line-following module uses a string of infrared emitters and detectors to follow a course. How do you know if an infrared LED is actually emitting infrared light? Your best bet is to use the camera in your mobile phone. Most cameras are at least partially sensitive to the infrared light wavelengths produced by IR LEDs (typically about 950 nanometers).

Activate the array and point the camera at it. Look at the scene from the digital viewfinder of the camera. If the IR LED is working, you'll see its light as white or pale blue. If the camera isn't picking up anything, do one last test to ensure it can see IR light. Point any infrared remote control at the camera and press some buttons. You should see a series of bright flashes.

BUILDING A LINE-FOLLOWING COURSE

Forget protractors, compasses, and straight edges. For a basic line-following course—good for practice and experimentation—just pull out a couple of sheets of poster board, and lay down some tape.

Using just a single sheet of poster board, begin by making a U-shape. Keep the line away from the edges of the board. As you apply the tape give it a bit of stretch . . . but not too much. The tape might contract as it springs back into its former size. You want to avoid excessively tight corners. If the edges of the tape pucker up in a curve, it's probably too tight.

On the same or another sheet of poster board, make a few sinusoidal curves, where the radius of the curves is the radius of the Teachbot's wheel base, or larger. When using spin turns (rather than slip turns), and with a sensor array like the QTR-8, you can expect the Teachbot to manage curves that are at least as small as its own wheel base diameter.

If you need a larger course, place two or more sheets of poster board end to end. Cut the tape off right at the edges of the paper. You don't need to overlap the tape for a perfectly seamless transition, but avoid gaps of more than 1/8".

Wanting to use something other than electrical tape? Feel free to experiment. But know that what looks black to you might be optically transparent to infrared light. Black marker is a good example. Unless you heavily apply marker ink to both sides of the paper, odds are it'll prove to be a poor choice for making line-following tracks.

If you have a laser printer you can experiment with making segments of tracks using a graphics program. Print out each sheet and tape them together. Be sure to use a fresh toner cartridge, and adjust the toner density to make it darker. The same technique may (or may not) work for inkjet printers. While pitch black to our human eyes, the ink is not opaque to infrared light.

Alternatives to black PVC electrical tape include black masking tape (try more than one layer thick), charting and graphics tape, and decal making tapes. While it's most common to use black tape on a white background, you can reverse the tones—just remember to also flip the programming logic in your sketch. Try using black construction paper or poster board and silver reflective tape.

Adding Wheel Encoders

Wheel encoders allow your robot to monitor the motion of its drive motors. Equipped with wheel encoders your robot can determine how far it's traveled, or whether its wheels are spinning at the same speed.

They're called *encoders* because they convert position or distance information of a wheel or other mechanical device to numerical data. The data output of an encoder can be absolute or incremental:

- An *absolute encoder* indicates the specific angular position of the encoder shaft, from 0 to 359 degrees. Once the shaft has turned a full 360 degrees, the angular position restarts at 0. This type of encoder is best used for such tasks as monitoring the position of a robotic leg or arm, so for the purposes of this chapter, this type isn't discussed here.
- An *incremental encoder* counts the number of transitions as the shaft turns. The total number of transitions for a full 360 degrees of rotation depends on the encoder—the count can range from as few as just one, to well over 1000 per revolution. Because they produce a simple pulse at each transition, incremental encoders are ideally suited for measuring travel distance and the speed of a robot's wheels.

The number of transitions per revolution determines the resolution of the encoder. The higher the transition count, the greater the resolution. The typical homebrew encoder has 4 to 18 transitions; commercially made encoders can have 64, 128, 256, 512, or even more transitions.

While accuracy is always an important aspect of any encoder system, there are definite advantages to keeping the resolution low, especially if you're making your own. The wheel encoders described in this chapter entail just six light/dark transitions, which is adequate for most tasks.

Depending on how you interpret the transitions, and given wheels with a diameter of 90 millimeters (3.54"), this equates to a positioning accuracy of up to 47 millimeters. How is this calculated?

- 90mm pi (3.14) = 282 mm — Distance traveled in one revolution of the wheel
- 6 transitions per revolution — Referred to as six *counts per revolution*, or 6 *cps*
- 6 / 282mm = 47mm — Travel distance between each transition

Later in this section you'll learn about various ways to increase encoder resolution.

TYPES OF WHEEL ENCODER MECHANISMS

Incremental wheel encoders employ numerous technologies to do their magic. There are three main types: optical, mechanical, and magnetic. In all cases the output of an encoder is a digital on/off pulse (or a varying voltage, but ultimately converted to a pulse). Internally, the encoder is composed of a disc, called the *codewheel,* and one or more *sensors* that read the changes as the codewheel spins.

Optical encoders use an infrared sensor and a series of alternating light/dark stripes on the codewheel. The codewheel can be either reflective (the sensor reads IR light that bounces off it) or transmissive (light passes through slots in the codewheel to a detector on the other side). Both types are shown in Figure 9-19. Optical encoders are among the least expensive and easiest to make at home, so they're the most common.

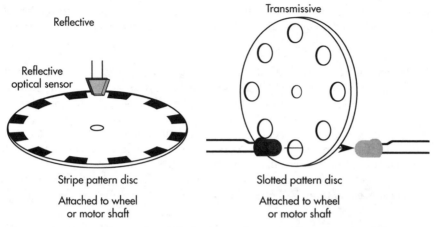

Figure 9-19 Two forms of optical encoders: reflective and transmissive. Of the two, reflective is generally easier to construct at home.

Mechanical encoders use a switch that opens and closes as the codewheel turns. They are typically used as digital volume controls. Because they are mechanical, they can wear out over time, and so they're not advised for use as wheel encoders on a robot.

Magnetic encoders use a metal gear as a codewheel, plus a Hall effect sensor, which is sensitive to magnetic fields. The sensor detects changes in the magnetic flux as the gear turns.

Encoders don't have to be attached to wheels to do their job, though this is the most common arrangement. One alternative is to connect the encoder directly to the shaft of the motor. The encoder may even be on the ungeared side of a double-shafted motor, in which case the encoder will turn much faster than the wheel on the other side.

Some encoder arrangements are unusual, but still useful: For example, you can attach sensors beside the painted treads of a tank-driven robot. As the treads turn, the sensors pick up the movement as the dark and light pattern pass by.

TEACHBOT ENCODER HARDWARE

Begin construction of the Teachbot's encoders by making a pair of disc inserts for inside of the wheels. The cutting and drilling pattern is shown in Figure 9-20. The exact diameter isn't critical, but it's better to err on the small side, so the disc will fit inside the rim of the wheel. I've placed mounting holes in the hub, which match holes already in the Pololu 60mm wheels specified for this project. However, for my prototype I didn't use the holes, and instead applied small bits of floral putty (available at most any craft store) to hold the disc inserts to the wheel.

Using a laser printer or plain paper copier (not an inkjet printer), create two codewheels using the pattern in Figure 9-21. For better quality, download the PDF version of this code-wheel from the ARB Support Site (*see* Appendix A). The pattern is made to be oversized, so that you can trim the codewheel to fit the disc inserts. After printing, punch a hole in the

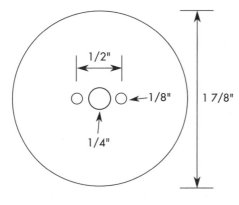

Figure 9-20 Layout for the disc insert. Construct using 1/8" plastic or aircraft-grade plywood.

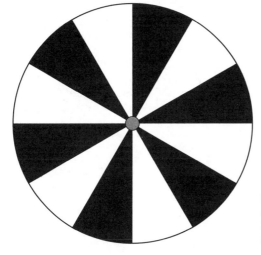

Figure 9-21 Codewheel with 6 black stripes (12 stripes total). See the ARB Support Site for a downloadable version you can print on your laser printer.

center of the codewheel. Apply the codewheel pattern over the disc using a glue stick or something similar. Use scissors or small hobby knife to cut away the excess paper around the disc. The encoder wheel assembly with codewheel pattern is shown in Figure 9-22.

When printing, adjust the toner setting so that the black stripes are as dark as possible. Be sure the black areas are *solid,* and not a mottled appearance, or the codewheel will not work properly. The inks in inkjet printers are usually not opaque enough to infrared light. If you only have an ink jet available, use it to print the codewheel, but take the printout to a copy shop and have duplicates made using a plain paper copier.

The electronics for the Teachbot DC wheel encoders consist of a pair of analog IR remitter/detector pairs. I used two Pololu QTR-1A (Pololu #958). They come with a trio of pin headers, which you need to solder onto the circuit board. The sensors are mounted on the bottom deck using miniature 3/8" L-angle brackets (Keystone #633, Mouser part number 534-633) and 2-56 machine screws and nuts, as shown in Figure 9-23.

Figure 9-22 Codewheel, disc insert, and 60mm wheel. For quick testing I'm using floral tape to stick the codewheel and insert to the wheel, but you can use fasteners for a more permanent job.

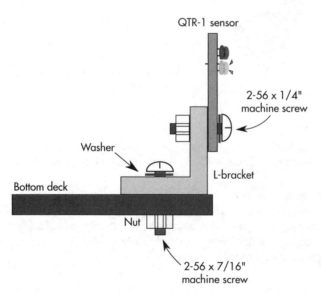

Figure 9-23 Assembly detail for the Pololu QTR-1A sensors. Use a small 3/8" metal bracket and 2-56 hardware.

Figure 9-24 QTR-1A sensor mounted beside the motor. When the wheel is attached, the emitter/detector pair of the sensor should be as close to the codewheel as possible without actually touching it.

Note that the hole in the bottom deck for the bracket is oversized for 2-56 hardware. This is to give you some variability in positioning the sensor. You want it as close to the codewheel as you can get it, without actually rubbing against it.

 Use flush cutters to trim the excess from the pin headers that protrude through the face of the sensor. Figure 9-24 shows the sensor mounted on the bracket. Note the slight tilt, to help line up the centerline of the sensor to the angle of the stripes.

 Use a 6" three-wire servo extension to connect the sensor to the breadboard. The QTR-1 sensors use the traditional Gnd-Power-Signal connection arrangement.

The QTR-1A provides an analog voltage relative to the amount of infrared light hitting the detector. One way to interface the sensor is to connect it to an Arduino's analog input. But all six of the Arduino's analog inputs are already occupied by the line-following module, and besides, using this type of connection is neither an efficient nor effective method of reading the fleeting pulses from an encoder.

Rather, the Teachbot uses a simple interface circuit that turns the analog voltage into a predictable LOW/HIGH pulse. See Figure 9-25 for wiring details. At the heart of the interface is an LM339 voltage comparator IC; the circuit uses two of the four comparators in the chip.

In operation, a trimmer potentiometer provides a reference voltage—something between 0 and 5 volts—to the comparator. This voltage represents the threshold between light and dark stripes on the codewheel. The circuit compares this threshold against the voltage provided by the sensor—the voltage is higher for a black stripe.

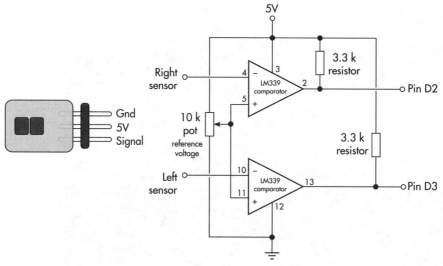

Figure 9-25 Interface electronics for the two infrared sensors. The voltage comparator changes the analog voltage from each sensor to a LOW/HIGH (off/on) signal. The potentiometer lets you set a reference voltage. Adjust this for best sensitivity from the sensors. If the outputs of the sensors are not well matched, add a second potentiometer so you can adjust each one individually.

On my Teachbot, white produced a voltage from the sensors of about 0.15V, and black about 3.65V. Yours should be fairly similar. Adjust the trimmer so that the voltage is between about 0.75V and 1.5V.

I'm using only one potentiometer to provide the reference voltage for both encoders. This is practical as long as the two encoders output more or less the same voltage range for white and black stripes. This is possible if the sensors are mounted identically, but electrical and physical variations may occur. If that's the case, you'll need to add separate pots, one for each encoder. Adjust the output of each pot to provide a clean LOW/HIGH signal transition at each stripe.

Use the following quickie sketch to check the levels from the two encoders. Temporarily connect the outputs of the QTR-1A sensors directly to analog pins A0 and A1, then run the sketch. Slowly turn the wheels and note the values. These values represent the voltages returned by the two sensors, and span the possible range of 0 to 1023. Lower values—ideally, as close to 0 as possible—will mean a white stripe; higher values, a black stripe. On my Teachbot, I get values of about 25 for white, and 780 for black.

```
void setup() {
  Serial.begin(9600);
}

void loop() {
  Serial.print(analogRead(A0), DEC);
  Serial.print(":");
  Serial.println(analogRead(A1), DEC);
  delay(250);
}
```

The output of the comparator is a nice and clean LOW and HIGH digital signal: it's LOW (or 0) when the sensor sees a black stripe; HIGH (or 1) for a white stripe. The signals from the comparator chip are fed to one of the Arduino's hardware interrupt pins, digital pins D2 and D3.

Why use these pins, and why use interrupts? The interrupt pins allow the Arduino to monitor the status of external hardware in the background. The main part of the loop in your sketch doesn't have to constantly check the current status of the external hardware.

Interrupts can be trigged by different logic levels; the *TeachbotDC_EncodersDemo* sketch demonstrates how to trigger the interrupts each time the Arduino sees a "rising" signal transition from LOW (black) to HIGH (white). Each time this interrupt occurs, the Arduino quickly branches off to an *interrupt handler* routine; inside this route is code to increase the encoder count by 1. As there are two encoders, each connected to a separate hardware interrupt pin, there are two interrupt handlers.

To use the sketch, compile and upload it. With the USB cable still connected, lift the Teachbot's wheels off the ground. Turn the motors on and open the Arduino IDE Serial Monitor window. After initializing, the motors will turn. You will see the current count values for the left and right motors, and may hear the speed of the motors change as the sketch attempts to allow the two motors to catch up to one another.

The *TeachbotDC_EncodersDemo* will run until the proper number of counts (40) has been registered for both wheels. At that time, the sketch exits the loop, and the motors stop. To rerun the sketch hit the reset button on the Arduino.

The interrupts are first established by associating them to an *interrupt number,* and the *name of the function* that serves as the interrupt handler routine. Interrupts are referred to by number, not by pin. In this way, differences in physical hardware won't affect your sketches. Just remember that in an Arduino Uno, interrupt 0 is for pin D2, and interrupt 1 is for pin D3.

So you know, the Arduino Mega 2560 provides additional hardware interrupts. This platform is recommended for more advanced robots, where you may need to interface many sensors to the Arduino.

TeachbotDC_EncodersDemo

```
#include <CompactQik2s9v1.h>

#include <SoftwareSerial.h>

#define rxPin 5
#define txPin 6
#define rstPin 7


SoftwareSerial mySerial =  SoftwareSerial(rxPin, txPin);
CompactQik2s9v1 motor = CompactQik2s9v1(&mySerial,rstPin);

int stopAt = 40;              // Go 40 "ticks" and stop
byte normalSpeed = 127;  // Regular speed
byte slowSpeed = 100;    // Governed speed

const int encRight = 2;
const int encLeft = 3;
int countRight = 0;
int countLeft = 0;
```

```
volatile int triggerLeft = 0;
volatile int triggerRight = 0;

void setup() {
  Serial.begin(115200);

  // Set pin modes
  pinMode(encLeft, INPUT);
  pinMode(encRight, INPUT);

  // Set up interrupts
  attachInterrupt(0, hitRight, RISING);   // Pin 2
  attachInterrupt(1, hitLeft, RISING);    // Pin 3

  mySerial.begin(9600);
  motor.begin();
  motor.stopBothMotors();
  delay(200);
  motor.motor0Forward(normalSpeed);     // Left motor
  motor.motor1Forward(normalSpeed);     // Right motor
}

void loop() {
  if (countLeft > stopAt) {
    motor.motor0Forward(0);
  }

  if (countRight > stopAt) {
    motor.motor1Forward(0);
  }

  if (countLeft >= stopAt && countRight >= stopAt) {
    motor.motor0Forward(0);
    motor.motor1Forward(0);
    return;
  }

  if (countRight >= countLeft) {
    motor.motor0Forward(slowSpeed);
    delayMicroseconds(15);
    motor.motor0Forward(normalSpeed);
  } else {
    motor.motor1Forward(slowSpeed);
    delayMicroseconds(15);
    motor.motor1Forward(normalSpeed);
  }

  // If left encoder hit, show results
  if (triggerLeft == HIGH) {
    Serial.print("L: ");
    Serial.println(countLeft, DEC);
    triggerLeft = LOW;
  }

  // If right encoder hit, show results
  if (triggerRight == HIGH) {
    Serial.print("R: ");
    Serial.println(countRight, DEC);
    triggerRight = LOW;
  }
}
```

```
// Interrupt handlers
void hitLeft() {
  static unsigned long last_interrupt_timeL = 0;
  unsigned long interrupt_time = millis();
  if (interrupt_time - last_interrupt_timeL > 15) {
    triggerLeft = HIGH;
    countLeft++;
  }
  last_interrupt_timeL = interrupt_time;
}

void hitRight() {
  static unsigned long last_interrupt_timeR = 0;
  unsigned long interrupt_time = millis();
  if (interrupt_time - last_interrupt_timeR > 15) {
    triggerRight = HIGH;
    countRight++;
  }
  last_interrupt_timeR = interrupt_time;
}
```

The TeachbotDC_EncodersDemo sketch is greatly simplified to help you see how the encoders work. It's not an ideal method for adjusting speed and reporting position. It is meant as a way to help you visually check that your encoders are working properly, and to understand how the interrupts and counting mechanisms work.

Refer to the ARB Support Site (see Appendix A) for additional examples of using encoders with the Teachbot.

ENHANCING ENCODER FUNCTIONALITY

The encoders on the Teachbot are relatively crude, and suffer from two ailments common to such hardware. You can somewhat compensate for them in code.

No direction sense There's no way to tell if the wheels are going clockwise or counter-clockwise. You need to surmise the robot's direction by checking against the current motor control function. In your sketches you'll need to add code to reset the encoder counters to 0 each time the robot stops, or decrement the counters when the wheels go in reverse.

Signal bounce Encoders tend to be "noisy" and may produce multiple pulses at each light/dark transition in the codewheel. The *TeachbotDC_EncodersDemo* sketch deals with this somewhat by including a debounce function in each interrupt handler. The code ignores pulses that occur within 15 milliseconds of one another—anything faster than this is probably a bounce. Adjust this value if you find the encoders seem to be either missing pulses or adding too many.

Programming code can only do so much. An advanced method for dealing with these issues is to use *quadrature encoders*. These not only count the number of pulses, but detect direction as well. When used with simple and low-cost interface electronics—which also include circuitry to condition the signal and remove noise—problems of signal bounce are all but eliminated.

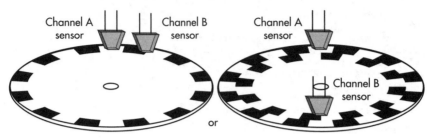

Figure 9-26   In quadrature encoding a second sensor is added so that it can read the codewheel pattern at a 90° phase.

Quadrature encoders use two stripes and/or two sensors, not just one, as shown in Figure 9-26. The stripes are oriented so that the pattern transitions are 90 degrees out of phase from one another, producing the following sequence:

LOW/LOW
LOW/HIGH
HIGH/LOW
HIGH/HIGH

…and repeat. The sequence of four is why it's called quad; *quad* as in four.

That's for an encoder going in one direction. The sequence is reversed when the encoder goes in the opposite direction. This is how a quadrature encoder knows which way it's going.

Quadrature encoding is an involved topic, and an important one. Sadly there's not enough space to do it justice here. I cover it in more detail in my book, *Robot Builder's Bonanza, Fourth Edition* (McGraw-Hill/TAB Electronics, 2011), with additional information provided on the ARB Support Site; refer to Appendix A for more information. The bonus projects on the Support Site include low-cost interface electronics and example Arduino sketches for use with quadrature encoders.

INCREASING ENCODER RESOLUTION

While wheel encoders with 6 cps are often more than enough for the average desktop robot, you may want higher resolution. Here are some ways to do it:

- *Register both the black and white stripes.* This provides for double the transitions—12 instead of 6, and therefore twice the resolution. This is accomplished by modifying the interrupt handlers in the *TeachbotDC_EncodersDemo* sketch to trigger on each signal change, rather than only on the rising edge of the signal.

```
attachInterrupt(0, hitRight, CHANGE);
attachInterrupt(1, hitLeft, CHANGE);
```

- *Use a disc layout with more stripes.* The practical number of stripes you can fit on the disc depends on the size of the sensing area—1/8" is typical—and the diameter of the wheel. The larger the wheel, the more stripes you can get on it. For any given wheel diameter, the more stripes you add, the thinner they have to be. If the stripes get too thin, the sensor can't discriminate between them. Ideally each stripe should be no thinner than the sensing area, or roughly 1/8".

● *Add a second sensor to create a quadrature encoder.* With the right interface electronics this can increase the effective resolution 2X or 4X. This arrangement can be difficult to implement using homebrew construction techniques, as the two sensors must be accurately positioned.

● *Replace your homebrew encoder with a commercial version.* These use better mechanics and optics, so resolution is improved; 32 cps+ isn't uncommon with even low-cost models. With quadrature encoding, the effective resolution of a 32 stripe unit is 128 cps.

Adding Sound Effects

As a bonus feature of the Teachbot DC consider adding a sound module to produce complex sound effects, including a mechanical-sounding voice. It's easiest to do this with a dedicated voice chip, and there are several to choose from. Options include the SpeakJet and the Babblebot—both share many of the same features, as they were originally developed by the same person.

For this project, I've picked the Babblebot sound generator, available as a stand-alone IC, as well as an Arduino shield and a separate breakout board; both the shield and breakout board contain a small audio amplifier. As the Teachbot DC doesn't use an Arduino that accepts shields, I elected to go with the breakout board version, which is shown already attached to the Teachbot in Figure 9-27. The board is mounted, using 1" nylon standoffs, to the right of the solderless breadboard.

Figure 9-27 Babblebot breakout board, mounted on the second deck of the Teachbot

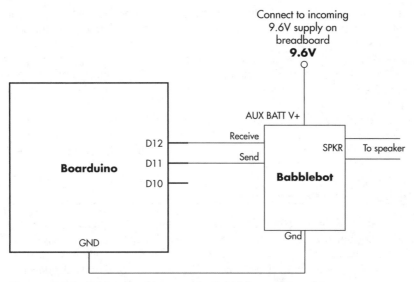

Figure 9-28   Wiring detail between the Babblebot and Boarduino

Electrical connection is simple, and is shown in Figure 9-28. The data cable is made using a pair of 3" precut wire with crimped female connectors. The Arduino end uses a 1 × 2 shell; the Babblebot end uses a 1 × 6 shell. Power for the board is derived from the *unregulated* 9.6V battery pack. The Babblebot board contains its own regulator for the Babblebot logic chip, and the onboard LM386 audio amplifier uses the higher voltage to produce a louder sound.

MOUNTING THE SPEAKER

The Babblebot board is mounted on standoffs, which provide room underneath to locate a slim-line 1" to 2" dynamic speaker. Generally speaking, the larger the speaker, the louder the sound. The speaker should have a 4Ω to 16Ω impedance.

I elected to go with a 0.5 watt 2.5" speaker, mounted on a third deck above (see Figure 9-29). This deck is like the top in the Teachbot Servo robot, described in Chapter 3. As shown in Figure 9-30, I'm using 2-1/2" long standoffs to separate the second and top decks, which provides ample room for components and wiring

The third deck is used to more advantage in the extended version of the Teachbot DC described in Chapter 11, "Going Places with the Telebot." In that version the additional deck provides space for a digital video camera, video transmitter and overlay module, a second Arduino, and various environmental sensors.

In this extended project you're able to drive your Teachbot DC using a remote control and see a real-time video image of the landscape ahead. On top of the video is overlay text, with current values from the environment sensors—temperature, humidity, compass direction, and more.

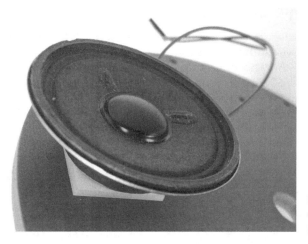

Figure 9-29 I mounted a dynamic speaker to the underside of a third deck.

Figure 9-30 Complete Teachbot DC, with third deck. This version of the robot will also be used in Chapter 11.

Whatever mounting you use:

1. Secure the speaker to the deck using a 1/2" square of double-sided foam tape.
2. Wire the speaker by cutting a 6" wire with crimped female connectors in half.
3. Strip off 1/4" of insulation from the cut end, and solder to the speaker terminals.
4. Insert the connectors in a 1 × 2 female shell. Be sure the connectors lock into place. Attach to the Speaker (SPKR) jumper pins on the Babblebot board.

PROGRAMMING THE BABBLEBOT

In addition to speech synthesis, Babblebot generates various types of sound effects, as well as music. It can even mix waveforms together to make complex sounds. The Babblebot uses a predefined library for the Arduino, called GinSing, available at *ginsingsound.com*. This library greatly simplifies using the Babblebot, and is open source.

Coding examples for the Babblebot would fill a book, so for the sake of simplicity, the example provided here demonstrates only core functionality. Be sure to download the GinSing library prior to trying out the examples. Unpack the library in the Arduino sketch *libraries* folder.

As with many open-source projects, the GinSing library is routinely updated, and changes may affect how the following example runs. One change you may need to make is the pins used to connect to the library. The Babblebot board wiring assumes the following pin connection:

Pin	Function
D10*	Overflow
D11	Send
D12	Receive

*In the Teachbot DC pin D10 is left unconnected, but is defined in sketches just the same. This pin is for the Babblebot chip's CTS line, which may not be implemented in all versions of Babblebot breakout boards.

As needed, modify the GinSing #define statements at the start of the sketch for these pins to match the wiring for the board. The default #define statement looks like this:

```
#define rcvPin  4
#define sndPin  3
#define ovfPin  2
```

Change to:

```
#define rcvPin  12
#define sndPin  11
#define ovfPin  10
```

As the GinSing library may change, it's best that you refer to the readme documentation for the details on which file(s) to modify.

The *babblebot* sketch demonstrates how to add simple sound effects to the Teachbot DC as it goes through its motion routines.

The *babblebot* sketch is too long to print here. Please download it from the ARB Support Site, where you will also find additional examples of producing sound effects.

Going Further with the Teachbot

There's even more to do with the Teachbot DC platform. Feel free to add the collision detection and avoidance sensors of your choosing. These sensors, and how they're wired to the Arduino, are covered in Part 1 for the Teachbot Servo.

Also be sure to read Chapter 11, "Going Places with the Telebot," where the Teachbot DC is remotely operated using a multifunction control. You have the option of adding video for a robot's-eye view as the bot explores its universe. And with a small collection of environment sensors, a second Arduino, and special video overlay hardware, you can teleoperate your Teachbot and get back data of its surroundings.

Chapter 10

The Amazing Tunebot

In the classic movie *Close Encounters of the Third Kind*, the (hopefully) friendly aliens reach out to us lowly humans with a series of five musical notes. We assume these five notes mean "yes we come in peace," but as Steven Spielberg never made a sequel to his film, for all we know they were really saying "we like earthlings medium rare."

Brain-eating space invaders notwithstanding, music has long served as a form of language. Music transcends generations and cultures. So, too, your robot can communicate using music. Previous chapters provided sound-making functionality using just the Arduino, or in the case of the enhanced Teachbot in Chapter 9, a voice box co-processor. But there's more. With a low-cost synthesizer board, you can turn your robot into an orchestra on wheels. Play single notes, chords, sound effects, even musical invitations to dinner with extraterrestrials.

In this chapter you'll learn how to build a robot with a repertoire of realistic-sounding instruments. I call the musically gifted robot *Tunebot*—see Figure 10-1 for what it looks like. Tunebot does more than just play music: it's actively *controlled* by the music you play on it. The robot demonstrates how playing individual notes on a piano-like infrared array actively directs the operation of Tunebot.

 Be sure to visit the ARB Support Site (see Appendix A) for additional code examples and variations. All sketches from this book are available for ready download, including those that are too long to fit in this book.

Figure 10-1 The completed Tunebot, ready to rumble

Tunebot Design Concept

Besides being fun to use, Tunebot's mission in life is to teach several important robotic concepts that you can use for other projects. As such, the robot is somewhat involved, so in turn this chapter is fairly long. Best to read it in a couple of sittings.

The design of the Tunebot includes the following nifty concepts:

- Traction over the ground using rubber tank treads, rather than the usual pair of wheels. Though the treads don't provide any specific advantages in this case—they're more helpful in robots used outdoors or over uneven terrain—they add a certain coolness factor.
- Locomotion via a pair of DC gear motors. These low-cost motors are assembled from a kit, where you can select the gearing ratio to select the overall speed of the robot.
- Motor management using a motor control circuit. The circuit lets you precisely drive the speed and direction of the motors.
- An all-in-one MIDI synthesizer board that plays musical notes and sound effects. You hear the sound through an amplified speaker that sits atop the robot.
- Novel eight-cell infrared array is used as an "air piano." Use it to control the Tunebot by playing notes. You hear the notes through the MIDI synthesizer as the robot obeys your hand and finger gestures over the sensor array.

- Sensors for directing the action of the Tunebot (bonus project). An ultrasonic ranging sensor and a mechanical leaf switch provide proximity and content detection.
- And, of course, an Arduino to supervise it all.

For the Tunebot I've departed from the usual "turtle" motif, just to spice things up a bit. That said, the body design is irrelevant to the function of the robot. You can make your Tunebot in any shape you want. Just keep in mind that robots with small rubber tracks can't be too large or heavy, or else the tracks will do nasty things like pop off in turns. I address the issue of detracking later in the chapter.

Building the Tunebot

The Tunebot body is composed of four main pieces: two sides, a back, and a bottom plate. The pieces are tied together using 3/4" × 3/4" plastic brackets and 4-40 hardware. Construction isn't particularly difficult, but things can get tight within the interior of the bot. For best results build and test the Tunebot in stages, as detailed throughout the chapter.

Begin by cutting and drilling out the parts as shown in Figure 10-2. All holes are 1/8". Hole location is moderately critical anywhere brackets are used, as the holes must match between back and sides. The width of the back and bottom should be the same. Otherwise, you can be fairly approximate in the other dimensions and hole locations.

Slots are used on the back to allow slipping the motors up and down, thereby tensioning the rubber treads. The slots are approximately 3/4" long and 1/8" wide. Make the slots by drilling two or three holes, then slicing through the holes with a small rat-tail file or coping saw blade. Don't make the slots too wide, or else the washers and nuts used to secure the motors may pop through. Try to make the slots as smooth as possible, so that the motor mounting screws slide through without binding up.

Begin construction by attaching the track idler rollers to the sides. These come with the Tamiya Track and Wheel set (Tamiya #70100). The set comes with three sizes of rollers: small, medium, and large. See Figure 10-3a for the hardware detail. Use 4-40 × 1" steel machine screws as "axles" for the rollers. Tighten the nylon locking nuts until the roller no longer turns freely, then back off one-quarter turn. Be sure the roller turns freely, but isn't so loose that it wobbles. Figure 10-3b indicates which rollers to use where.

Use 4-40 × 1/2" machine screws and nuts to mount six plastic brackets to the back and sides, as depicted in Figure 10-4a and b. Four are mounted on the inside back; two on the outside. Notice that the screws and nuts for the lower set are shared by the brackets on the inside and outside.

ASSEMBLE AND ATTACH MOTORS

Tunebot uses two Tamiya #70093 three-speed crank axle gearbox motor kits. To construct the motors you'll need a #0 Phillips screwdriver and small needle-nose pliers, plus a (included) hex wrench. A pair of flush cutters—like the kind for snipping off the ends of wires when constructing circuit boards—is useful for separating the various plastic pieces.

Tunebot's gearboxes can be constructed using one of three speed ratios: 17:1, 58:1, and 204:1. The 17:1 speed is too fast for the Tunebot, so opt for either the 58:1 or 204:1 ratios. The higher the ratio, the slower the robot, but the more power it has.

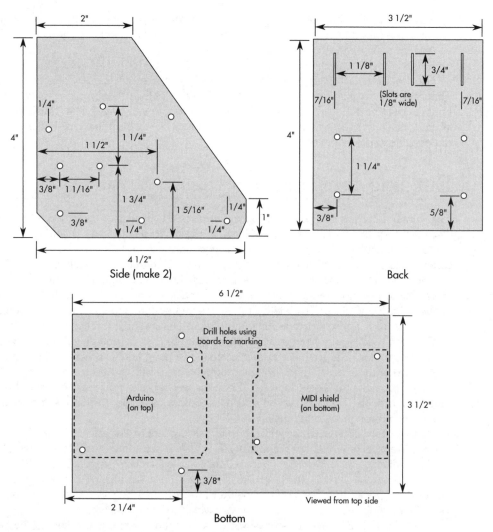

Figure 10-2 Drilling and cutting guide for the Tunebot body parts. Hole placement is critical only for those used to mate with a plastic bracket.

I recommend starting with the 58:1 speed. In any case, if you want to try a different speed you can always disassemble the motor and rebuild it at the new setting. Be sure to save all the extra parts in a reclosable plastic sandwich baggie so you can change out the gearing should you ever want to.

Each gearbox kit comes with a long hex-shaped drive shaft. You'll need to cut the shaft to length as shown in Figure 10-5. Do this before final assembly of the gearbox.

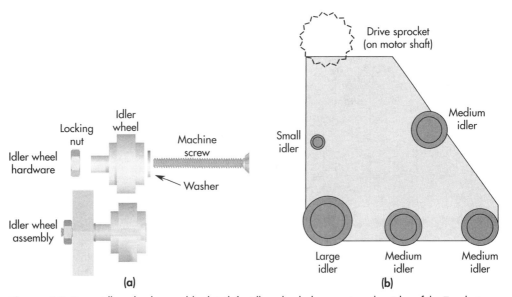

Figure 10-3 **a.** Idler wheel assembly detail. **b.** Idler wheel placement on the sides of the Tunebot. Use the small, medium, and large idlers as shown.

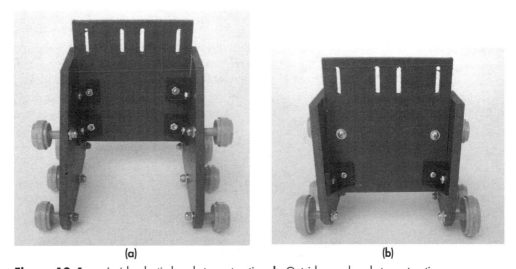

Figure 10-4 **a.** Inside plastic bracket construction. **b.** Outside rear bracket construction.

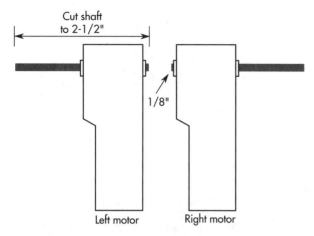

Figure 10-5 Cut the shafts for each motor to 2-1/2" and adjust lengths as shown. You will make a left and a right motor.

After each gearbox has been built (but before you have inserted the small DC motor into the gearbox), loosen the setscrew that holds the drive shaft in place. Adjust the position of the shaft within the gearbox as shown in the illustration. You want about 1-3/8" of the shaft protruding from the outside of the gearbox. Retighten—but don't overtighten!—the setscrew when the shaft is in the proper position.

(In order to access the setscrew you may need to manually rotate the gearing to make the screw accessible. Each motor comes with its own miniature hex wrench for tightening the setscrew. Be sure to save the wrench for use later.)

Figure 10-6 shows the two gearboxes assembled and with the large drive sprocket (from the Track and Wheel set) attached. The miniature DC motor has not yet been inserted into the gearbox.

Mount the two motors to the back of the Tunebot using four 4-40 × 7/16" flathead screws and nuts. The length of these screws is important. You don't want to use screws that are too long, or they'll protrude into the inside of the Tunebot cavity and interfere with the placement of the battery.

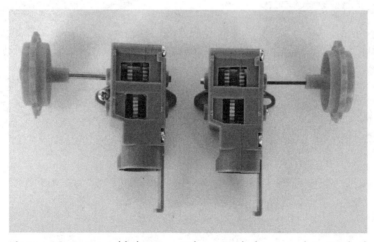

Figure 10-6 Assembled motor gearboxes, with drive sprockets attached

ATTACH BOTTOM

Following Figure 10-7, use 4-40 × 1/2" machine screws and nuts to attach a pair of plastic corner brackets to the left and right sides of the Tunebot. Slide the bottom between the side pieces, and secure it to the brackets using 4-40 screw hardware.

ASSEMBLE AND ATTACH TREADS

Assemble the treads from the Tamiya Track and Wheel set using all of the individual lengths of the track material. Each track will have four segments. The segments interlock into one

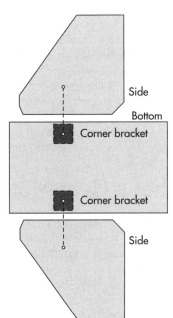

Figure 10-7 Plastic bracket construction to attach the bottom base to the sides

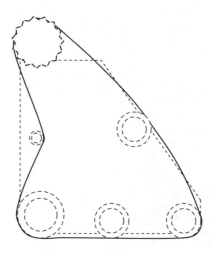

Figure 10-8   The "running" arrangement for the treads. When not using the Tunebot, unhook the tread from the small idler wheel in the back. This releases tension from the rubber and keeps it from overstretching.

another. Use only your fingers; avoid using mechanical force or tools to link the segments together, or else the rubber may get torn. It can take a few tries to get the hang of it. Refer to the instructions that come with the set for assembly details.

To install the tread, loop over the four idler wheels. Wrap the tread around the teeth of the drive sprocket, and carefully push the sprocket onto the motor shaft. Tension the tread by looping it inside the small idler wheel at the rear of the bot. Figure 10-8 shows the treading used when operating the robot. We'll call this the "running" position of the tread.

The design of the Tunebot allows for adjusting the tension of the track—important, because if the treads are too loose, they'll easily pop off. With the treads in the running position, loosen the screws to the motors, and slide them up or down so there is modest tension in the rubber. The treads should not be too loose or they'll pop off. If they're too tight, operation of the Tunebot may suffer, and the treads may stretch out of shape. You'll need to experiment until you find a happy medium. This part takes some patience and practice.

After the motor position has been set, unhook the treads from their rear idler wheels. This releases the tension to the treads and prevents them from stretching when the robot is not in use. If you don't do this, over time the rubber will "ease out" and you'll need to readjust the motors again to take up the slack. As the rubber in the treads is not reinforced, it's possible for it to become so overstretched that adjusting the motors no longer tensions them. You'll need to replace the treads with a fresh set.

This completes the basic Tunebot construction. You are now able to wire and test the motors and complete its electronic subsystems. Let's turn first to the Tunebot's motors.

Testing and Using the Tunebot's Gear Motors

DC motors come in various types. The most common is the permanent magnet motor, which (in its typical form) is composed of a spinning rotor surrounded by a set of magnets. The rotor is composed of numerous windings through which current flows. Put simply, the movement of current through these windings is what makes the rotor spin.

A chief feature of most permanent magnet motors is that they are reversible (see Figure 10-9): Connect a battery to the two terminals of the motor and the motor spins in one direction. Reverse the polarity of the battery and the motor turns in the opposite direction.

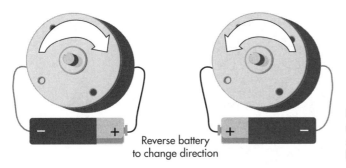

Figure 10-9 Most permanent magnet motors can be reversed simply by switching the direction of current from a battery or other power source.

Reverse battery to change direction

Obviously, for a robot you don't want to keep pulling the batteries out and reconnecting them the other way just to change directions. The most straightforward way to demonstrate motor control and reversing is with a switch. So let's try that. The section that follows is optional, but recommended if you'd like to learn more about how the Tunebot's twin motors maneuver the robot. You'll create a small control panel using small switches that let you manually steer the Tunebot around a room.

MAKING A CONTROL PANEL

Construct a simple control switch panel using a small piece of 1/8" thick plywood or plastic sheet. The switches provide wired control of the motors; by flipping the switches forward and back you can experiment with how the motors make the Tunebot travel forward and back and make turns.

You'll need:

2	Double-pole, double-throw center off momentary switches
4	6-32 × 1" machine screws
4	6-32 nuts
4	6-32 acorn (cap) nuts
1	Cable clamp (1/4")
1	3-cell AA battery holder
Misc	20- or 22-gauge stranded conductor hookup wire, telephone wire (4 wires inside)

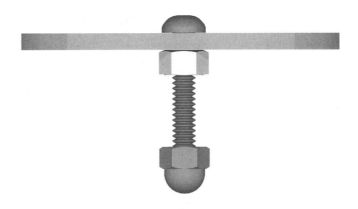

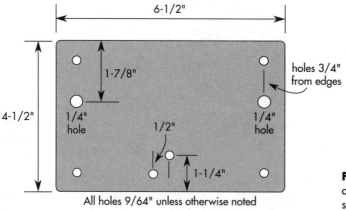

Figure 10-10 Cutting and drilling guide for the (optional) switch control panel

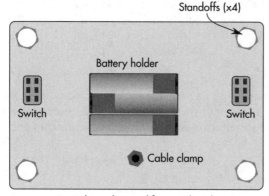

Control panel viewed from underside

Figure 10-11 Placement of switches, battery holder, and standoff legs on the underside of the control panel

Refer to Figure 10-10 for the drilling and cutting template for the control panel. Holes are 9/64" unless noted. The switches attach to the control panel using their own threaded nuts. Use one or two 1" squares of Velcro to secure the battery holder to the underside of the panel. Figure 10-11 shows where to attach the switches, corner standoffs, and cable clamp on the control panel.

WIRING THE MOTORS AND CONTROL PANEL

A set of two double-pole, double-throw (DPDT) switches operate the motors and show how differential steering works. You can practice using the switches to see how different switch positions affect the motion of your robot.

For a full appreciation of how to control the Tunebot, choose miniature DPDT toggle switches that have a center-off position, as well as momentary spring-loaded "on" action. In electronics catalogs and online shopping carts, you'll often see these listed as:

(on)-off-(on)

or

(mom)-off-(mom)

where the (on) or (mom) in parentheses means that the switch is spring loaded. Releasing the toggle returns the switch to its off position.

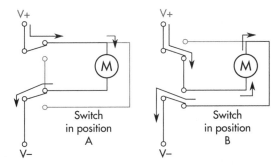

Figure 10-12 Concept of using double-pole, double-throw (DPDT) switches to reverse the direction of a motor

Each switch controls one motor. In the center-off position, the motor is unpowered, so it doesn't move. The motor turns in one direction when the switch is pushed toward either on position. With the switch pushed toward the other position, the motor turns in the opposite direction.

Figure 10-12 shows the current path from battery through the motor, depending on the position of the switch. The darker lines represent the current path. Note that in one position of the switch, current from the battery flows through the motor one way; when the switch is in the other position, current from the battery flows through the motor in the opposite way. This is what causes the motor to reverse directions. The wiring is either "straight through" or crossed. When crossed, the polarity is reversed, so the motor is reversed.

 In a robot like the Tunebot you need two switches, one for each motor. The way each motor spins determines the direction of the robot. This is the same basic technique used in the Teachbot covered in earlier chapters. Depending on the current direction of the two switches, the robot travels forward or backward or spins right or left.

Refer to Figure 10-13 for how to wire the switches on the control panel to the motors on the Tunebot.

1. Start first by connecting the battery holder to the switches. Each switch needs power from the batteries. Be sure not to cross up the red (positive) and black (negative) connections between the switches, or you might create a short circuit.
2. Wire the criss-cross of the switches as shown. This criss-cross is what makes the motor reversal work.
3. Wire the switches to the motors using an 8- or 10-foot length of four-conductor modular telephone extension. Clip off the modular plugs on either end, and strip back about five inches of the outer insulating jacket to expose the wires inside. The wires are color coded, but it doesn't matter which actual colors you use for connecting between the switches and motors. Solder the switch and motor connections as shown.

 On the control panel end, thread the wire through the extra hole before soldering to the switches. This provides added strain relief.

Telephone wiring is made to be very flexible, which is why it's chosen here. Many kinds have a thin filament that when soldered burns away. It takes practice getting a good solder job with this stuff, so take your time and try again if your first attempt doesn't turn out. If you

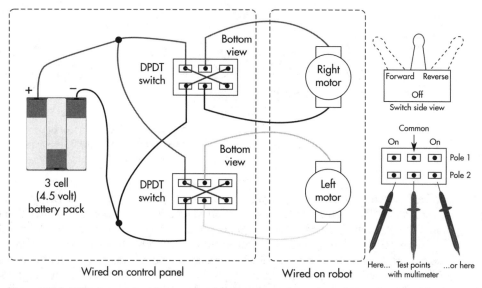

Figure 10-13 Connection diagram for the control panel switches. Wiring from the panel to the Tunebot is through a length of standard telephone cable (they're flexible and contain the requisite four wires).

continue to have trouble, try a different brand of wiring, or substitute with four-strand ribbon cable. The ribbon cable isn't quite as flexible, but it will do for our purposes. Avoid the use of solid conductor wire.

The terminals on virtually all DPDT switches provide for the wiring diagram as shown in Figure 10-13. The switch has six terminals, two rows of three terminals each. The "Common" connections of the switch are the two terminals in the middle.

If your switch came with a wiring diagram, check it to be sure the two center terminals are Common. If you have a multimeter, dial it to test continuity, and apply one test lead to a Common terminal and the other lead to a terminal on either side. Toggle the switch one way, then the other. In one of the positions the meter should show continuity.

Keep the telephone wiring from pulling out by making a small coil around a 1/4" U-shaped plastic cable clamp. The clamp acts as a strain relief. Use a clamp on both the control panel and the robot base.

With the wires attached to the motors, fit the motors into their gearboxes. The solder terminals should be oriented so that they are not blocked by the body of the gearbox. The motors "click" into place. The design of the gearbox allows you to readily remove the motor—we'll do exactly that later in the chapter; see "Replacing the Tamiya Gearbox Motors" for details.

A wired and finished Tunebot control panel is shown in Figure 10-14. Note the use of three AA batteries. This provides the Tunebot motors with about a nominal 4.5 volts when using non-rechargeable cells. The stock motors that come with the Tamiya gear box kits are intended for use at 3 volts, so they'll run a little fast. That's okay for now.

Figure 10-14 Underside of completed control panel, wired and ready to go

MANUALLY OPERATING THE TUNEBOT

Review your handiwork to check for accidental shorts, bad soldering joints, or other problems. When you are satisfied all looks good, insert three AA batteries (standard alkaline or rechargeable) into the battery holder. Temporarily remove the tracks from the Tunebot, then try each of the control switches. The motors should turn as you flip the switches.

Figure 10-15 shows operating the two switches on the control panel. Test the wiring between the switches and motors by pressing both switches forward (away from you).

- *The robot moves forward when you press both switches forward.* The motors and switches are connected with the correct polarity. You're ready to keep playing!

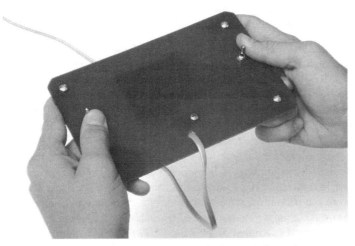

Figure 10-15 Use your thumbs to toggle the control panel motor switches forward and back.

- *The robot moves backward.* Loosen the nuts on the switches, and rotate both 180 degrees. Retighten the switch nuts.
- *The robot turns in a circle.* Loosen the nut on the switch controlling the motor that's going in the wrong direction. Rotate the switch 180 degrees, then retighten the nut.
- *The right and left switches control the wrong motor.* Flip the control panel around. Or, remove both switches, reverse their position on the panel, and put them back in.

Place the Tunebot in the center of the room. Practice steering the robot around the floor. You'll note that tight, spinning turns are performed by pulling back one switch while the other is pushed forward.

A slower "pivoting" turn is accomplished by releasing one of the switches (depowering its motor) and letting the other motor continue. But this operation is *not* recommended for the Tunebot, due to its rubber tracks. Single-side turns can cause extra stress on the tracks, causing them to pop off.

To negotiate a turn, *always* use so-called tank steering where one track moves forward while the other moves backward. This will help reduce throwing tracks.

Using Electronic Motor Control

As you discovered in the previous section, motor direction is controlled by alternating the polarity of the current applied to the motor's terminals. Mechanical switches help demonstrate how to control the motors of a robot, but they are manual devices. In an autonomous robot you must rely on some electronic means to turn its motors on and off and change direction.

Through electronic control, you can replace the switches with circuitry. This circuitry effectively duplicates the action of those DPDT switches. And because the circuitry can be operated electrically, other electronic components can be connected as sensors to provide automatic function.

MOTOR CONTROL USING AN H-BRIDGE

An *H-bridge circuit* is the most common way to provide all-electronic control of a motor. It's called an H-bridge because the schematic diagram for such a circuit depicts an "H" pattern in the wiring. Figure 10-16 shows a simplified H-bridge (don't try to build this circuit; it's just for demonstration).

In operation, only two transistors (in opposite corners) are ever turned on at the same time. (If all transistors are turned off, the motor receives no current, and so it stops spinning.) The direction of the motor is altered by switching which two transistors are activated.

While there are many—and I mean *many*—workable H-bridge designs, most of the ones you can easily replicate leave something to be desired. They rely on big and bulky transistors, so the finished circuit is ungainly. And to make the thing easier to construct at home, some useful components are often omitted, which can affect its efficiency and dependability.

I find it easier (and usually cheaper!) to get a ready-made H-bridge. A popular one is the L298, a complete H-bridge in a single integrated circuit. One L298 will operate *two* motors. The L298 comes in a special kind of high-wattage package that isn't designed for breadboarding, and it needs some external parts—specifically a set of protection diodes—to complete its circuitry.

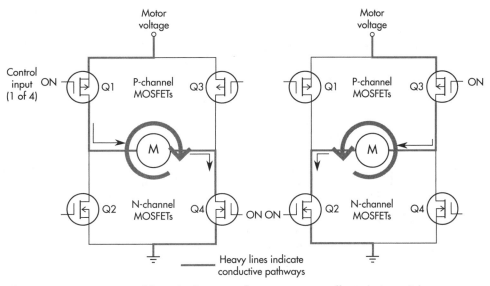

Figure 10-16 Concept of the H-bridge, using four transistors to effectively "route" the current through a motor. The direction of current through the motor causes it to go in one direction or the other.

For the sake of convenience, I like to use the many ready-made L298 circuit board modules available. Most are priced at between $20 and $35, depending on features. The one I've specified for the Tunebot is the ArduMoto shield from SparkFun. It uses a surface mount version of the L298, and incorporates those all-important protection diodes.

You're free to use another Arduino motor shield, as long as it has similar features as the ArduMoto. The most important is the current-carrying ability of the bridge. It must be at least two amps *per motor*. Most any motor shield based on the L298 module will share this specification.

Note, too, the four pins for controlling the two motors. If your shield uses different pins for the motor functions you'll need to adjust the sketches in this chapter accordingly. Likewise, changes in the motor control pins will upset the pin assignments for the Tunebot's "piano keyboard" and sensors.

As a shield, the ArduMoto fits atop the Arduino. The board comes *sans* header pins; you'll need to solder them on. You'll also need to solder three terminal blocks to the front of the board for connecting the battery and two motors. Some variations of the ArduMoto don't include the pin headers or terminal blocks, so you'll have to get these separately if you get one of the bare-board versions.

Replacing the Tamiya Gearbox Motors

The small DC motor that's included with the Tamiya 3-speed gearbox is rated for operation at just three volts. And as motors go, it's not terribly efficient. As a result, these motors consume a lot of current, especially if they get bogged down under lots of weight or strain.

All H-bridge circuits are rated to handle a certain amount of current before they overheat—and either shut off, fail, or become damaged. The SparkFun ArduMoto motor control shield specified for the Tunebot uses an L298 motor bridge module, which can handle up to two amps per motor. Yet the Tamiya motors can consume in excess of three or four amps when powered at just four to six volts.

Fortunately, you can get low-cost drop-in replacements for the stock motors provided in the Tamiya gearbox kits. For example, the Pololu #1117 replacement motor is rated for operation at 6 volts (can be operated at up to 12 volts), and at that voltage consumes a maximum of only 800 mA (0.8 amps) when stalled (stalled means the motor is still powered but cannot move).

When replacing the motor you'll need to pull the small spur gear off the old motor, and mount it onto the replacement. The fit is tight; avoid grabbing the teeth of the gear with a tool to remove it. Instead place the jaws of a pair of small needle-nose pliers behind the gear, and gently work the gear off. Be careful that the gear doesn't come flying off the motor shaft, or you may never find it!

Solder 7" leads (use 20 to 22 AWG stranded wire) to the motor for connecting it to the L298 module. Here's a tip: It's a good idea to add a small 0.1 μF ceramic disc capacitor directly between the terminals of the motor, along with the connecting wires. The capacitor helps to reduce any negative effects of electrical noise induced by the motor. Be sure the capacitor is rated at 25 volts or higher.

Mounting the Arduino and Other Electronics

You're now ready to attach the Arduino and other electronics to the Tunebot. Start by removing the bottom plate so you have better access to it. You'll find it easier to remove the screws holding the bottom to the corner bracket, rather than removing the brackets as well.

MIDI BOARD

Solder right-angle headers to a SparkFun Musical Instrument Shield (DEV-10587), then attach the shield to the underside of the bottom plate, as shown in Figure 10-17. Use 4-40 × 7/16" flathead screws and nuts. Thread the screws from the top, so that the nuts are on the side of the board. Add a couple of plastic (not metal) washers under the board as spacers.

So you've noticed I'm using the SparkFun Musical Instrument Shield as a separate breakout board, rather than as an Arduino shield. Why? Two reasons: First, the shield uses pin 3 to transmit data to the MIDI chip on the board. That pin is already used by the ArduMoto. While it's possible to modify the shield so that it uses a different pin for communications, I opted not to for the sake of simplicity.

And second, motor H-bridge circuits are known for producing lots of electrical noise—actually, it's not the H-bridge that makes the noise; it's the motors. Nevertheless, by placing the MIDI board separately, there's less chance of electrical noise bleeding into the audio circuits and causing unpleasant buzzing and crackling noises.

Figure 10-17 SparkFun MIDI shield, shown with right-angle headers on its power, signal, and audio output terminals. It's mounted on the underside of the bottom base, rather than used as a standard Arduino shield.

Make some connecting cables to connect between the MIDI board and Arduino. You'll need:

2   2-wire cables, 12" in length. Terminate both ends with 1×2 female pin headers.
1   2-wire cable, 3" in length. Terminate only one end of the pair with 1×1 female headers.

Refer to Figure 10-18 for an interface circuit to go between the output of the MIDI board and a 1/8" female audio jack. You can solder the resistors and capacitor directly onto the

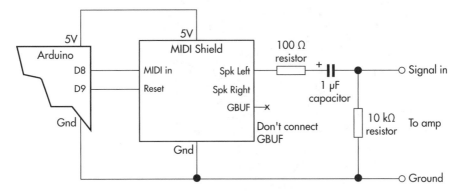

Figure 10-18 Hookup diagram for the MIDI shield. The interface circuit allows you to connect an external amplifier and speaker to either left or right channel. Important! Leave the GBUF terminal unconnected.

terminals of the audio jack (that's what I did), or you can construct the circuit on a small board. Use double-sided foam tape to mount the board on the rear corner of the bottom base. A small self-contained amplified capsule speaker plugs into the jack.

Note that the circuit only uses one side of the stereo output of the MIDI board. This simplifies construction. While the MIDI board outputs in stereo, the sketches for the Tunebot don't use the right and left channels separately. So you can tap off from either left or right channel to just one speaker connection.

Do not connect anything to the GBUF terminal on the board. This terminal is *not* a ground connection. Rather, it's a constant voltage reference (about 1.23V) for use when listening to the MIDI board through small headphones. Leave GBUF unconnected, and use one of the GND terminals on the MIDI board for its electrical connection to the 1/8" audio jack.

Drill some 1/4" or 3/8" feed-thru holes into the bottom to route the MIDI cables to the top side. Drill one hole near the MIDI board power terminals, and drill two more toward the front of the bottom plate. The hole beside the power terminals is for the audio output wires; the holes at the front of the plate are for MIDI power and signal cables.

Reattach the bottom plate to the sides of the Tunebot. Figure 10-19 shows the MIDI board wired and secured to the bottom plate.

ARDUINO

The Arduino is attached to the front end of the bottom base using a set of 4-40 × 7/16" flathead screws, 1/8" nylon spacers, and 4-40 nylon nuts (instead of nylon nuts, you can use

Figure 10-19   The MIDI board attached and wired on the bottom of the Tunebot

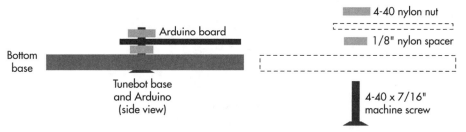

Figure 10-20 Use 1/8" nylon spacers to mount the Arduino board to the bottom base. The power plug and USB connector should face forward for easy access. (The power plug isn't used in this project, but you'll need a way to insert the USB cable to program the Arduino.)

metal nuts with #4 plastic washers as insulators). Thread the screws up from the bottom, carefully align the spacers, then secure the Arduino using the nuts. See Figure 10-20 for details.

The Tunebot bottom base is drilled to use the lower-left and upper-right mounting holes on the Arduino (USB and power jack on the left side). Only two screws are needed to hold the Arduino in place, as there's no weight or stress put onto the board.

ARDUMOTO SHIELD

The SparkFun ArduMoto is a standard Arduino shield. Get the version with the power and motor terminal blocks, plus shield headers. You need to solder these parts onto the shield prior to use.

As with all shields, the ArduMoto fits over the Arduino through mating pins located on the bottom of the board. It is *extremely* important that you install the ArduMoto correctly, or else the shield and even the Arduino may quickly burn out. The reason: Misaligning the pins between Arduino and shield may cause a serious short.

I had this happen to me when building the first Tunebot prototype. I was too eager to try things out, and didn't notice that the lower pins—the ones carrying power through the two boards—were shifted one to the right. The power pins of the Arduino are laid out in such a way that if the pins are shifted by one in either direction a short will always occur. So be careful out there!

CONNECTING MOTORS TO MOTOR SHIELD

Locate the two motor terminal blocks on the side of the shield. The motor terminals are labeled A1/A2 and B1/B2. Strip about 1/4" from the ends of the motor wires, and lightly tin with solder (to keep the wire strands together). Attach the wires from the left motor to the A1/A2 terminals, and the wires from the right motor to the B1/B2 terminal. Use a small #0 Phillips screwdriver to tighten the terminals so the wires remain snug.

While you're at it, prepare the power wiring for the system as follows:

1. Strip 1/4" insulation off a 5" length of 18-gauge stranded red jacketed wire. Tin the ends with solder.

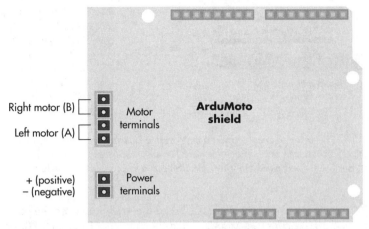

Figure 10-21 Terminal connections on the SparkFun ArduMoto H-bridge motor shield. Be sure to screw terminals (don't solder) so you can readily connect or change the wiring to the motors.

2. Do the same for a black jacketed wire.
3. Attach the wires to the power terminals on the ArduMoto shield as shown in Figure 10-21. The red wire goes to the + terminal; the black wire goes to the – terminal.

PROTOSHIELD

A standard Arduino solderless breadboard prototyping shield provides a fast and convenient way of connecting the MIDI board, sensors, and other parts to the Arduino (the ArduMoto is already attached, as it's connected as a shield). You can use most any prototyping shield that conforms to the following basic specifications:

- Standard shield layout, with extended male/female header pins for Arduino I/O
- Mini solderless breadboard (17 column by 10 row)
- Reset button exposed on top, to allow resetting Arduino when needed

Before attaching the protoboard to the Arduino/ArduMoto stack, prepare the solderless breadboard as shown in Figure 10-22a. Parts consist only of a 1 µF tantalum capacitor, jumper wires, and double-length male header pins—the latter allow you to connect wires with female header connectors into the breadboard.

The breadboard layout includes the position of 2- and 3-pin double-long male header pins, used to connect to the wiring for the MIDI board, plus the bonus project leaf switch and Ping ultrasonic sensor. If you don't plan on adding the ultrasonic sensor or switch, you can leave out these parts. You can always add them in later.

With the protoshield complete, carefully position it over the motor shield, being careful not to misinsert any of the pins. At a minimum you'll need to connect the MIDI power and signal pins as shown in Figure 10-22b—you can leave the other wires and components off the board

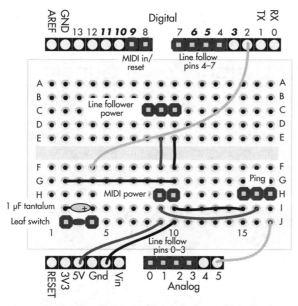

Figure 10-22   a. Component and wiring guide for the solderless breadboard on the prototyping shield. The dark gray octagonal blocks indicate double-long male header pins, used to connect to wires with female headers.

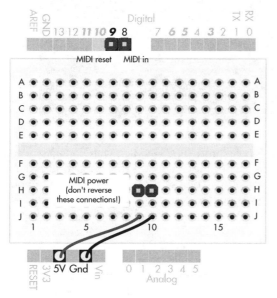

Figure 10-22   b. Detail of the MIDI board connections on the prototyping shield

for now. *As with all electrical connections to circuitry, be very careful that you do not reverse the power leads to the MIDI board.* Otherwise, things go poof.

Constructing the Battery Power Supply

Tunebot derives its power for the motors and electronics from a single 9.6V rechargeable battery pack. I'm using a set of eight AA-size NiMH cells in a shrink-wrapped pack. The pack terminates in a Tamiya polarized battery connector; this connector type is not critical, as long as it's polarized. You'll need to remove the batteries from time to time to recharge them, and you don't want to accidentally connect them backwards. Doing so may cause irreversible damage to all the electronics on your Tunebot.

I used a pushbutton single-pole, single-throw (SPST) switch with contacts rated at about 3A, and mounted it in a hole on the side of the Tunebot. Press once to turn the robot on; press again to turn it off. The style of switch you use for your Tunebot isn't important, but strive for one that isn't so big it crowds out the rest of the components. Avoid bare solder contacts on the back side of the switch. Apply tape or heat-shrink tubing as needed to prevent accidental shorts.

Wire the switch and battery as shown in Figure 10-23. I didn't use a fuse on my prototype Tunebot, but it's not a bad idea to include one. A small inline fuse rated no less than 2A should suffice. If the fuse blows when working the Tunebot's motors, opt for a slightly higher rating. Take note that only the positive (red) lead of the battery goes through the switch. The black lead goes directly from the battery disconnect to the negative (–) terminal on the ArduMoto. Solder all wires, and apply heat-shrink tubing to insulate any bare connections.

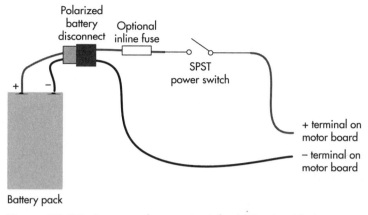

Figure 10-23 Battery and power circuit for the Tunebot. The battery should be 9.6V rechargeable AA size (or equivalent). Use a quick-disconnect so you can easily remove the battery for recharging.

Power and Motor Tests

With power and electronics connected, you're ready to upload a test sketch to the Arduino.

1. Temporarily remove the treads from their sprockets and idlers. This allows you to observe the action of the motors without the robot going anywhere.
2. Connect a freshly charged battery pack, and turn on the power switch.
3. Connect the Arduino to your PC, enter the following *Tunebot_MotorForward* sketch, and upload it to the Arduino.

The sketch starts automatically. Assuming the motors have been connected properly, the motors should alternately start and stop.

Turn the main power switch off, and put the treads back on. Wrap the treads in front of the small idler to put the robot in its "running" mode. Place the robot on the ground, disconnect the USB programming cable, and turn the main power switch back on.

Observe the action of the motors. If one or both motors turn in the wrong direction—the robot goes backwards or turns—remove power and flip the terminal wiring from the affected motor(s) on the H-bridge.

Tunebot_MotorForward

```
int pwm_a = 3;              // Left motor
int dir_a = 12;

int pwm_b = 11;             // Right motor
int dir_b = 13;

void setup() {
  setupDrive();
  Serial.begin(9600);
```

```
    delay(500);
    digitalWrite(dir_a, LOW);   //Set motor direction forward
    digitalWrite(dir_b, LOW);
    analogWrite(pwm_a, 255);    //Set motor speed (full on)
    analogWrite(pwm_b, 255);
}

void loop() {
    //empty
}

void setupDrive() {
    pinMode(pwm_a, OUTPUT);    //Set control pins to be outputs
    pinMode(pwm_b, OUTPUT);
    pinMode(dir_a, OUTPUT);
    pinMode(dir_b, OUTPUT);
}
```

Next try a more complete motion test using the *Tunebot_MotorExercise* sketch. Repeat the previous process (remove treads, apply power to the robot, upload sketch, replace treads). This test changes the direction and even the speed of the two motors, providing a more thorough test.

Tunebot_MotorExercise

```
int pwm_a = 3;              // Left motor
int dir_a = 12;

int pwm_b = 11;             // Right motor
int dir_b = 13;

void setup() {
    setupDrive();
    Serial.begin(9600);
}

void loop() {
    drive_fwd();            // Drive forward, delay 2 seconds
        delay(2000);
    drive_turnR();
        delay(2000);
    drive_turnL();
        delay(2000);
    drive_rev();
        delay(2000);
}

void setupDrive() {
    pinMode(pwm_a, OUTPUT);    //Set control pins to be outputs
    pinMode(pwm_b, OUTPUT);
    pinMode(dir_a, OUTPUT);
    pinMode(dir_b, OUTPUT);
}

void drive_fwd() {
    digitalWrite(dir_a, LOW);
    digitalWrite(dir_b, LOW);
```

```
    analogWrite(pwm_a, 100);  // Full speed (100%)
    analogWrite(pwm_b, 100);
}

void drive_turnR() {
  digitalWrite(dir_a, LOW);
  digitalWrite(dir_b, HIGH);
  setPWM(pwm_a, 70);          //Set speed to 70%
  setPWM(pwm_b, 70);
}

void drive_turnL() {
  digitalWrite(dir_a, HIGH);
  digitalWrite(dir_b, LOW);
  setPWM(pwm_a, 85);          //Set speed to 85%
  setPWM(pwm_b, 85);
}

void drive_rev() {
  digitalWrite(dir_a, HIGH);
  digitalWrite(dir_b, HIGH);
  analogWrite(pwm_a, 100);
  analogWrite(pwm_b, 100);
}

// Set PWM speed using percentages,
//  map to 0 to 255 (255=fastest)
void setPWM(int pin, int val) {
  analogWrite(pin, map(val, 0, 100, 0, 255));
}
```

Do the treads come off when the Tunebot is moving or turning? They may not be tight enough. With the treads in the "running" position, try retensioning the treads by loosening the motor screws and pushing the motors a little higher. Tighten the screws and try again.

Rubber treads are sensitive to the type of surface they are operated on. Avoid thicker carpets or polished floors. A basic kitchen tile flooring or very low nap (like that in offices or schools) is best.

HARDENING THE DRIVETRAIN

The replacement motors and high torque gearing of the Tunebot drivetrain put a lot of stress on the plastic drive sprocket. The hex-shaped shaft can (and eventually will!) strip out the insides of the drive sprocket hub. You can prevent/repair this by adding a 1/4" (inside diameter) metal Dura-collar around the hub of the sprocket.

Place the collar around the hub (it'll be a tight fit), then use the included setscrew to tighten it against the plastic. Variations in molding may cause the hub to be too large for the collar. You can try enlarging it with a drill (the metal underneath is brass, so it's fairly soft), or temporarily expanding it in a small pot of boiling water. Heat for a couple of minutes, then use a pair of pliers—don't try holding it with your fingers!—to fit it over the hub. Wait until the metal completely cools before inserting and tightening the setscrew.

Attaching Line Follower Module

Tunebot uses a commercially made line follower module from Parallax, shown already attached to the robot in Figure 10-24. It's mounted on the back of the Tunebot and oriented so that its array of detectors point straight up. Before attaching the line follower board to the robot, let's take a closer look at what it does and how it works.

The Parallax line following module contains a series of eight infrared emitters and eight detectors. The emitters and detectors use light to measure proximity to the module. With the emitter/detector pairs the module functions as an eight-note (one octave) piano keyboard. Waving your hand or fingers over the IR detectors triggers a note to play through the Tunebot's MIDI sound module.

To resist being influenced by other light sources, the infrared emitters on the line following module are modulated at a frequency of 38 kHz to 43 kHz. The detectors are tuned to only respond to light in this frequency range. The same technique is used with infrared remote controls; in fact, the IR detectors used on the module are the same kind found in remote control appliances like DVDs and TVs. Rows of (non-connected) female headers are placed between each sensor to prevent cross-channel interference.

A small multi-turn trimmer potentiometer on the module alters the modulation frequency. By adjusting the pot you can change the sensitivity—and therefore the effective range—of the sensors. The change from one frequency extreme to another is not great. You'll get the best sensitivity with the pot dialed in at about middle, which is the center frequency response of the IR detectors.

Each of the eight detectors has its own output pin. When a detector senses a nearby object (light from the LED emitter is reflected back into the detector), its pin goes LOW. A small LED on the other side of the module lights up to visually indicate when a detector has been triggered.

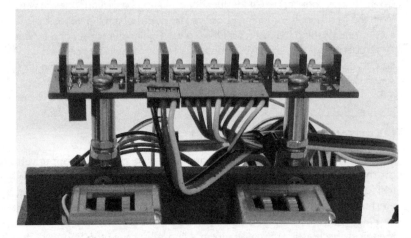

Dark colored objects don't reflect infrared light as well as light colored ones. Wearing black cotton gloves will yield poor results, as the fabric absorbs the light, rather than reflects it. For a unique sound style, try playing the infrared keyboard using various household items—ice cream sticks, (light colored) hair combs, paper with black and white markings on it, sparkling Christmas ornaments, whatever suits your fancy.

Figure 10-24 The Parallax line following module acts as an "air piano" to control the Tunebot using finger and hand gestures. The MIDI board produces musical notes as you wave your fingers over the sensor module.

With the theory of operation out of the way, let's now turn to attaching the line follower module on the back of the Tunebot. The module comes with 1" metal standoffs; use these with small metal L-brackets to mount the board above the motors on the rear of the robot. You'll need to first remove the motors. Attach the brackets to the back with 4-40 × 3/8" flathead screws. The screw heads must be flush against the plastic so that they won't protrude and interfere with the motors.

Once the line follower has been attached you can replace and retension the motors.

Making Music with MIDI

As noted at the start of this chapter, Tunebot plays its music through a MIDI synthesizer board. The board is located on the underside of the robot. A small amplified capsule speaker is connected to the audio output of the MIDI board. The arrangement is shown in Figure 10-25. Before programming the Tunebot to wander and make music, let's explore a bit more about MIDI and how it works.

The speaker placement shown in Figure 10-25 is for when the Tunebot is not rolling around. During actual play, you will need to relocate the speaker into the insides of the robot. Why? The motors produce a lot of electrical noise, and many (though not all) capsule amps are designed to mute the speaker output if the noise level gets too high. This is so you don't get an earful of static. By simply moving the speaker away from the motors you can avoid the sound cutting in and out as the Tunebot goes about its merry way.

Figure 10-25 Use a self-contained capsule amplifier/speaker to connect to the MIDI board. In the lower-left corner you can see the 1/8" audio jack; the resistors and capacitor interface circuit in Figure 10-18 are directly soldered to the terminals of the jack. Keep the speaker in place with Velcro. Note: When operating the Tunebot move the speaker to the inside of the robot, away from the motors.

MIDI stands for *Musical Instrument Digital Interface.* It's a standard method for controlling electronic instruments—it does other jobs, too, but music is what we're interested in here. The MIDI specification covers the data transmission itself, the electrical connection, even the hardware used to link everything together.

For a self-contained robot we're mainly interested in the data talking part. MIDI speaks by sending short *messages* over an asynchronous serial connection.

- The data sender is referred to as a *controller.* A common MIDI controller is an electronic keyboard, but there are many other kinds—for the Tunebot the controller is an Arduino.
- The data receiver is referred to by various names, such as *sound module, synthesizer,* and *sound bank.* Its job is to listen for commands sent by the controller and turn them into musical notes. The sound module is connected to an amplifier and speaker so you can hear the music.

MIDI message

0xB0	0x07	127
8-bit command byte	7-bit data byte	7-bit data byte

Figure 10-26 MIDI messages are composed of just a few bytes, an eight-bit command byte (it always has its eighth bit set to 1), and one or more seven-bit data bytes.

Most messages are only two or three bytes long. Each message starts with an eight-bit *command* (or *status*) byte, followed by one or more seven-bit *data* (or *parameter*) bytes. The combination of command and data byte(s) of a single message is an *event.* Figure 10-26 shows a simplified example of a three-byte MIDI message. By starting every message with an eight-bit byte, and then only using seven-bit data bytes, the sound module can more easily keep sync between itself and the controller.

As a programming practice, command bytes are usually entered in code in *hexadecimal* (hex) format. On the Arduino, hex values are preceded with a *0x* prefix. For example, *0xB0* is the same as decimal *176*. Data bytes are entered as decimal or hex values, whichever method you prefer. I'll provide some examples shortly.

OF INSTRUMENTS, BANKS, AND CHANNELS

The most basic MIDI module is programmed to reproduce the sound of at least one instrument. But many are capable of synthesizing numerous instruments—acoustic and electric piano, guitars, various woodwind and brass instruments, drums, bongos, cymbals, and much more.

Sound modules have built-in defaults, and for instruments the default is usually an acoustic grand piano. To change the instrument so the module plays something else, you send the "change instrument command," *0xC0,* followed by the number of the instrument you want to use. Instrument numbers span from 0 to 127, with 0 being the default.

For instance, to change to instrument #12—whatever it might be—you simply transmit a couple of bytes from the controller to the MIDI sound module:

```
0xC0 12
```

(Before going further take note that the documentation for the sound module probably lists the instruments starting at 1, but programmatically you specify them beginning with 0. Just keep that in mind when you meant to choose "Voice Oohs" and instead get "Choir Aahs.")

Recall that seven-bit data bytes are limited to just 128 values. Many modern MIDI modules are capable of reproducing more than 128 instruments, including all the various drum and special effects sounds. The number of instruments can be increased by separating them into two or more *banks*. With up to 128 banks, and 128 instruments per bank, you could conceivably choose from 16,384 different instruments. There may be few MIDI modules with that many instruments programmed into it, but the potential is there.

As shown in Figure 10-27, instruments are played through *channels;* each channel can sound a different instrument. MIDI supports 16 channels. Many (but not all) command bytes are defined as *channel messages.* The eight bits of the command byte are divided into two four-bit sections (see Figure 10-28):

- The first four bits specify the *command,* such as changing instruments.
- The second four bits indicate the MIDI *channel,* from 0 to 15 (16 channels total).

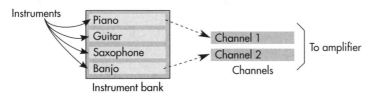

Figure 10-27 MIDI devices synthesize one or more instruments that are kept in a repository called a bank. Instruments are played through any of 16 channels.

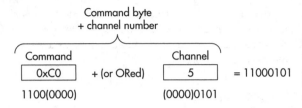

Figure 10-28 Channel voice bytes are constructed by adding or ORing the four most significant bits (MSB) of a command byte with the four least significant bits (LSB) of the channel number.

For example, the command byte for switching instruments (*0xC0*) begins with the binary bits *1100*. Channel_0 is binary *0000*, making the whole command byte:

```
1100 0000
```

Other variations:

```
11000001     Channel_1
11000010     Channel_2
11000011     Channel_3
```

and so on. To combine a command byte with a channel number, simply add them together. Popular programming practice performs this function using *binary OR*, which on the Arduino is done using the | (pipe) operator:

```
0xC0|3        // Bits: 11000011
```

As with instruments, channel numbers in MIDI are typically noted as 1–16, but programmatically are referred to as 0–15. So in the preceding list, *Channel_0* is actually MIDI channel 1, *Channel_1* is MIDI channel 2, and so on.

This difference really only matters if you're interfacing with other MIDI equipment, like a MIDI keyboard, where channels are numbered 1 through 16. For programming using an Arduino and a MIDI breakout board like the one in Tunebot, the channel offset numbers make little or no difference.

UNDERSTANDING CHANNEL VOICE MESSAGES

Switching instruments to play something else is an example of modifying the channel voice. *Channel voice messages* are those that alter the sound on a specific channel rather than for the entire sound module. There are seven channel message commands in common use.

Command	Binary*	Hexadecimal
Note Off	1000*nnnn*	0×80
Note On	1001*nnnn*	0×90
Polyphonic Key Pressure	1010*nnnn*	0×A0
Control Change	1011*nnnn*	0×B0
Program Change	1100*nnnn*	0×C0
Channel Pressure	1101*nnnn*	0×D0
Pitch Bend Change	1110*nnnn*	0×E0

* The *nnnn* bits specify the channel (0 to 15).

Of these, the following four are used in Tunebot:

- **Program Change**   Selects which instrument to play
- **Note On**   Turns on a note of a specific *pitch* and *velocity* (more about these terms next)
- **Note Off**   Turns off the note
- **Control Change**   Selects a variety of operations, such as changing the volume of the channel or selecting a different bank of instruments

Each note you play on a MIDI device has two primary characteristics, pitch and velocity. Pitch is the frequency of the note, and velocity is its volume. On a piano, each white or black key on the keyboard delivers a different pitch. How hard and quickly you press down on the key determines its volume, or velocity.

MIDI uses a simple numbering system to denote pitch; these numbers (see Figure 10-29) match the white and black keys on a piano. Given a standard 88-key piano keyboard, the MIDI pitch numbers vary from 21 (low tone) to 108 (high tone). The sequential numbering counts both the white and black keys. Very low and very high tones are possible by extending the pitch values through the full 0–127 range. For reference, middle-C on the piano is MIDI pitch value 60. It has a pitch (sound frequency) of 261.6 Hz (Hertz, cycles per second).

For note velocity, the values range from 0 (silent) to 127 (full volume). There is also a separate channel volume command (see next) that sets the overall level for the whole channel. Values likewise range from 0 to 127 for channel volume. The example sketches in this chapter demonstrate setting both note velocity and channel volume.

An example *Note On* message might look like this:

```
0x90|1 69 100
```

- *0x90* is the *Note On* command byte; added (ORed) to it is 1, for channel_1.
- *69* is the pitch of the note. The value 69 represents the A above middle-C.
- *100* is the note velocity. A value of 100 is about 3/4 full volume (full = 127).

Note that I'm showing only the command byte as hexadecimal, and the rest are plain ol' decimal.

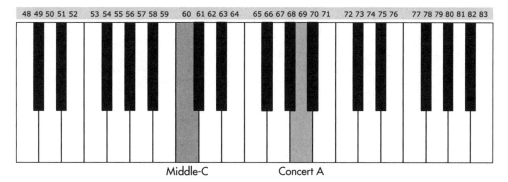

Figure 10-29   MIDI pitches are defined using a standard numbering system that corresponds to the white and black keys on a keyboard.

UNDERSTANDING CONTROL CHANGE MESSAGES

The *control change message* allows for an extended range of channel functions. These messages use the *0xB0* command byte with a channel number, a control byte, and a data byte to set the value. The three most common control change messages—and the ones used in the example sketches—are:

Option	Control Byte
Bank Select	0x00
Channel Volume	0x07
All Notes Off	0x7B

Examples:

```
0xB0|0  0x00  0x78  // Sets bank 0x78 (decimal 120) on channel 0
0xB0|5  0x07  63           // Sets volume to 63 on channel 5
0xB0|3  0x7B  0            // Turns off all notes playing on channel 3
```

MAKING TUNES

The best place to start with MIDI is creating a single tone. Sketch *MIDI_VerySimple* demonstrates basic functionality of making MIDI sounds from the Arduino. The sketch, which is designed for Arduino 1.0, begins with including the *SoftwareSerial* object library (it comes with the Arduino IDE). The library creates an object named *MIDI*, which is set to transmit data on pin D3.

MIDI_VerySimple

```
#include <SoftwareSerial.h>
SoftwareSerial MIDI(255, 8); // Rx, Tx (Rx not used, Tx pin=3)

void setup() {
  MIDI.begin(31250);          // Set up serial comm to MIDI
  resetMidi(9);               // Reset MIDI device (pin 8)
  ctrlMIDI(0, 0x07, 127);     // Set channel volume to full
  ctrlMIDI(0, 0x00, 0x79);    // Select 'melodic' bank
  sendMIDI(0xC0|0, 0, 0);     // Select instrument in bank
}

void loop() {
  noteOn(0, 60, 127);         // Middle-C note, velocity=127
  delay(1000);                // Wait one second
  noteOff(0, 60,  127);       // Turn off note
  delay(1000);
}

// Turn note on (press key)
void noteOn(byte channel, byte note, byte attackVelo) {
  sendMIDI( (0x90 | channel), note, attackVelo);
}
```

```
// Turn note off (release key)
void noteOff(byte channel, byte note, byte releaseVelo) {
  sendMIDI( (0x80 | channel), note, releaseVelo);
}

// Set controller value with channel
void ctrlMIDI(byte channel, byte data1, byte data2) {
  sendMIDI( (0xB0 | channel), data1, data2);
}

// Send bytes to MIDI device
void sendMIDI(byte cmd, byte data1, byte data2) {
  MIDI.write(cmd);
  MIDI.write(data1);
  // Pass only if valid 2nd byte
  if( (cmd & 0xF0) <= 0xB0 || (cmd & 0xF0) == 0xE0 )
    MIDI.write(data2);
}

// Reset MIDI device
void resetMidi(int resetPin) {
  pinMode(resetPin, OUTPUT);
  digitalWrite(resetPin, LOW);
  delay(100);
  digitalWrite(resetPin, HIGH);
  delay(100);
}
```

In order to try this sketch you'll need to attach the amplified speaker to the interfaced output at the rear of the Tunebot (refer to Figure 10-18). Or, for testing purposes, you can insert a pair of headphones into the headphone jack on the MIDI board.

Let's take a look at how the sketch works: The *setup()* function starts the MIDI serial link at 31250 baud, then resets the board by toggling the Reset pin LOW, then HIGH.

The remainder of the *setup* function sends a series of messages to the MIDI board. These messages are transmitted with the assistance of some helper functions, *ctrlMIDI* and *send-MIDI,* found later in the sketch:

- *ctrlMIDI* is for passing control change messages. The function takes three arguments: the channel number, the specific control to change, and the value to change it to. Example: The line *ctrlMIDI(0, 0x07, 127)* alters the volume on channel 0 to *127*.
- *sendMIDI* is for passing generic messages. This function also takes three arguments: the command byte and up to two data bytes. The statement line *sendMIDI(0xC0 | 0, 0, 0)* sets the instrument on channel_0 to *0*.

All of the boards from SparkFun use a synthesizer chip with the standard General MIDI instrument 1 (GM1) set. In these instruments 0 is an acoustic grand piano. The MIDI shield chips offer two instrument banks, referenced as *0x79* for a set of "melodic" instruments and *0x78* for a set of drum and other percussion instruments. If you don't specify a bank, the chip will default to bank 0, which is the same as the melodic bank.

Take a closer look at the *sendMIDI* function itself. Notice that there's a bit of code to determine if the message contains two or three bytes:

```
if( (cmd & 0xF0) <= 0xB0 || (cmd & 0xF0) == 0xE0 )
```

The logic is this: If the most significant four bits—the ones on the left—are equal to or lesser than *0xB0, or* equal to *0xE0,* the message is assumed to have three bytes. If the test doesn't meet these requirements, the message is assumed to have just two bytes. The *send-MIDI* function takes three bytes regardless, so when passing a two-byte message just fill in the third parameter with a 0.

The *loop()* function cycles through turning on and off a note at one-second intervals. Two function calls are used here: *noteOn* and *noteOff.* They're simply wrappers to the *sendMIDI* function, and make the coding simpler. Both take the same arguments, but they do different things:

- *noteOn* turns on a note of a specific pitch and at a specific velocity (loudness).
- *noteOff* turns the note off. You want a *noteOff* for *every* *noteOn;* otherwise, the notes will just congeal into one another. (This continues until the MIDI chip runs out of internal resource space.)

The first argument of the *noteOn/noteOff* methods is the channel. I'm using channel 0, but there's nothing stopping you from using another channel. The second argument is the pitch—in this case, it's middle-C on the piano. The third argument is the velocity. I'm using the highest value possible, 127.

USING SELF-DESCRIPTIVE CONSTANTS

Programming MIDI is made easier by using constants instead of raw hexadecimal or decimal values. Rather than trying to remember that *0xC0* switches instruments, you can use a constant like *CTRL_INSTRUMENT,* which mnemonically helps you to remember what the value does.

Constants are defined in an Arduino sketch using either of two methods:

```
const int SOME_VALUE = 5;
```

or

```
#define SOME_VALUE 5
```

Both do the same thing: they define a *symbol* that evaluates to a number. When the sketch is compiled, the Arduino substitutes each instance of *SOME_VALUE* with the number 5. Constants are similar to variables, in that both let you reference a value by name. But unlike variables, constant definitions don't take up memory in the Arduino. Take notice that when using the *#define* statement there's no semicolon at the end of the line.

The *MIDI_Simple* sketch demonstrates the same functionality as the previous sketch, but uses constants to refer to the command and data bytes. While the sketch is longer, it takes up the same amount of program space inside the wee innards of the Arduino.

MIDI_Simple

```
// Channel voice messages (all are 8-bit, 128 and above)
#define   NOTE_OFF        0x80    // 2 bytes: key, velocity
#define   NOTE_ON         0x90    // 2 bytes: key, velocity
#define   CONTROLLER      0xB0    // 2 bytes: controller num, value

// Control change messages
#define   CTRL_BANK       0x00    // Select bank of instruments
#define   CTRL_VOLUME     0x07    // Set volume (0-127)
#define   CTRL_INSTRUMENT 0xC0    // Set instrument within bank
#define   CTRL_ALL_OFF    0x7B    // All notes off

// Instrument banks
#define   BANK_MELODIC    0x79    // "Melody" instruments
#define   BANK_PERCUSSION 0x78    // Percussion (drum) instruments

// General values
#define   Instrument      5       // Instrument: electric piano 2
#define   NOTE_C4         60      // Middle-C on piano
#define   NOTE_A4         69      // A above middle-C (concert pitch A)
#define   CHAN_1          0       // Channel (1 of 16)

// Pin assignments
#define   midiIn          8       // MIDI input
#define   resetPin        9       // Reset

byte m_velocity = 127;            // Note velocity (volume)

#include <SoftwareSerial.h>
SoftwareSerial MIDI(255, midiIn); // Rx, Tx (Rx not used)

void setup() {
  MIDI.begin(31250);              // Set up serial comm to MIDI
  resetMidi();                    // Reset MIDI device
}

void loop() {
  //set channel volume
  ctrlMIDI(CHAN_1, CTRL_VOLUME, m_velocity);

  //select channel bank
  ctrlMIDI(CHAN_1, CTRL_BANK, BANK_MELODIC);

  //select instrument in bank
  sendMIDI(CTRL_INSTRUMENT, Instrument, 0);

  //turn on note
  noteOn(CHAN_1, NOTE_C4, m_velocity);
  delay(1000);

  //turn off note
  noteOff(CHAN_1, NOTE_C4,  m_velocity);
  delay(1000);
}

// -- Standard functions --
// Turn note on (press key)
void noteOn(byte channel, byte note, byte attackVelo) {
  sendMIDI( (NOTE_ON | channel), note, attackVelo);
}
```

```
// Turn note off (release key)
void noteOff(byte channel, byte note, byte releaseVelo) {
  sendMIDI( (NOTE_OFF | channel), note, releaseVelo);
}

// Set controller value with channel
void ctrlMIDI(byte channel, byte data1, byte data2) {
  sendMIDI( (CONTROLLER | channel), data1, data2);
}

// Send bytes to MIDI device
void sendMIDI(byte cmd, byte data1, byte data2) {
  MIDI.write(cmd);
  MIDI.write(data1);
  // Pass only if valid 2nd byte
  if( (cmd & 0xF0) <= 0xB0 || (cmd & 0xF0) == 0xE0 )
    MIDI.write(data2);
}

// Reset MIDI device
void resetMidi() {
  pinMode(resetPin, OUTPUT);
  digitalWrite(resetPin, LOW);
  delay(100);
  digitalWrite(resetPin, HIGH);
  delay(100);
}
```

PLAYING MULTIPLE NOTES

Nearly all MIDI synthesizers are *polyphonic,* meaning they can play more than one note at a time (conversely, *monophonic* means one note at a time). You can play multiple notes on the same channel or on a different channel. If the notes are from the same instrument, it's easier to use the same channel. Simply repeat the *noteOn* statement for each pitch you wish to play.

Conversely, if you wish to play notes from different instruments, you'll need to specify an instrument on each channel. Then call *noteOn* for each note, referencing the channel you wish to use.

MIDI_Chords shows how to use multiple notes to create chords. The sketch also shows cycling through a series of instruments, playing a full octave's worth (12 steps) for each instrument. You can use this program to sample the various instruments to discover how they sound.

Like the *MIDI_Simple* sketch, *MIDI_Chords* also uses constants to define numeric values. Only this time they're placed in separate header files to make it easier to manage and update the code. These header files are kept in the same folder as the sketch, and are referenced at the top of the sketch using *#include* statements.

MIDI_Chords

```
#include <SoftwareSerial.h>
#include "instruments.h"
#include "constants.h"

byte m_bank = BANK_MELODIC;    // Bank select (melodic)
byte m_instrument = 0;         // Instrument in bank (melodic only)
byte m_note = 0;               // Note (key, pitch)
byte m_velocity = 127;         // Note velocity (volume)

byte minKey = NOTE_C4;         // Beginning pitch for demo
byte maxKey = NOTE_C4 + 12;    // Ending 1 octave higher
```

```
// Starting and ending instruments
byte minInstrument = Acoustic_Grand_Piano ;
byte maxInstrument = Rock_Organ ;

// Rx, Tx (Rx not used)
SoftwareSerial MIDI(255, midiIn);

void setup() {
  Serial.begin(9600);
  MIDI.begin(31250);          // Set up serial comm to MIDI
  resetMidi();                // Reset MIDI device
}

void loop() {
  playBasic();                // Continuous loop
}

// Main routine, plays block of pitches (notes)
//   within a selected instrument range
void playBasic() {
  //set channel volume
  ctrlMIDI(CHAN_1, CTRL_VOLUME, m_velocity);

  //select channel bank
  ctrlMIDI(CHAN_1, CTRL_BANK, m_bank);

  // Iterate through instrument(s) to play
  for(m_instrument = minInstrument ;
   m_instrument <= maxInstrument ;
   m_instrument++) {

    // Select instrument within bank
    sendMIDI(CTRL_INSTRUMENT | CHAN_1, m_instrument, 0);

    // Play through notes
    for (m_note = minKey ; m_note <= maxKey ; m_note++) {
      noteOn(CHAN_1, m_note, m_velocity);
      noteOn(CHAN_1, m_note+4, m_velocity);
      noteOn(CHAN_1, m_note+7, m_velocity);
      delay(300);

      //turn off just 1 note at a time
      //noteOff(CHAN_1, m_note,  m_velocity);

      //turn off channel sounds
      ctrlMIDI(CHAN_1, CTRL_ALL_OFF, 0);
      delay(100);
    }

    delay(150);               // Pause before next instrument
  }
}

// Remainder of Standard functions code
```

 Important! MIDI_Chords will not compile without the extra header files that are defined at the start of the sketch. Rather than reprint them here, you can find them ready for download at the ARB Support Site (Appendix A).

The multiple notes are played using this code:

```
noteOn(CHAN_1, m_note, m_velocity);
noteOn(CHAN_1, m_note+4, m_velocity);
noteOn(CHAN_1, m_note+7, m_velocity);
```

The *m_note* variable contains the root note to play. The additional two notes are defined as *m_note+4* and *m_note+7*; these numeric intervals create a major chord (remember: the MIDI pitch values count half-steps; that is, both the white and the black keys). The trio of notes are contained in a *for* loop, which cycles through a series of 12 half-step notes.

For every note that's turned on, you need to eventually turn it off. This can be done by simply repeating the pitch with a *noteOff* statement, or you can use a special "All Notes Off" command to remove all notes playing on the channel. The *MIDI_Chords* sketch uses the latter method to demonstrate how it's done.

Play around with the delay after the notes are turned on. Some instruments require a longer *attack*—the amount of time for the note to come to full volume, timbre, and form. If the delay is too short, the notes will literally sound cut off.

Programming Robot Motions and Music

It's time to put the Tunebot hardware, MIDI, and Parallax line follower module to work. Refer to *MIDI_Piano* for a working demonstration. The program listing is too lengthy to reproduce here; you may download it ready to go from the ARB Support Site.

Begin by wiring the line follower to the Arduino as shown in Figure 10-30. The module comes with the cabling; all you need to do is attach it to the Arduino I/O pins and breadboard using double-long male header pins.

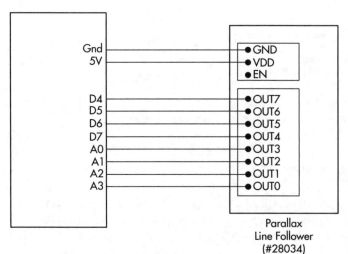

Parallax
Line Follower
(#28034)

Figure 10-30 Wiring diagram between the Arduino and the Parallax line follower module

Simple movement patterns over the line follower module control the bot:

- Pass your fingers over the center sensors to make the robot go forward. Activate the center sensors again and the robot reverses direction.
- Activate the rightmost sensors to turn right; the leftmost sensors to turn left.
- Cover all the sensors (or at least the first and last) to make the robot stop.

Experiment with the line-sensor coding to add other finger-gesture controls. For example, you might add code that listens for a specific three- or four-note tune in order to activate the Tunebot. Create some code to play your favorite songs while the Tunebot is moving.

Extending the Tunebot with Proximity and Touch Sensors

With a couple of basic sensors you can have your Tunebot play music while it wanders a room, with sound effects that trigger in response to any obstacles that are encountered. Figure 10-31 shows a Parallax Ping ultrasonic sensor and a standard leaf switch attached to a faceplate in front of the Tunebot.

This faceplate is mounted to the robot using some quick and easy extensions (made with 1/8" plastic) and 3/4" nylon threaded standoffs. You can follow the same general layout for the faceplate for your bot, or create something new and unique.

Figure 10-32 shows the electrical connection for the switch; refer to Figure 10-22a for the connection point on the breadboard for the Ping sensor.

Figure 10-31   A Ping ultrasonic sensor and leaf switch—bent outward to extend its reach—are optional features for the Tunebot to enable it to detect obstacles.

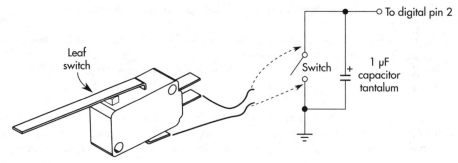

Figure 10-32   Wiring diagram for the leaf switch. A 1 μF capacitor across the switch terminals acts as a debouncer circuit.

 As with *MIDI_Piano*, the sketch for running the Tunebot with ultrasonic sensor and touch switch is too lengthy for inclusion here, and to save space it may instead be found on the ARB Support Site (see Appendix A).

On the Web: Enhancing the Tunebot with Rigid Tracks

Also on the ARB Support Site is a bonus project where you'll learn how to improve the track system of the Tunebot robot using rigid treads. These treads, available commercially from a number of online retail sources, are made of hard plastic, so they won't pop off as easily as the all-rubber Tamiya treads.

The downside is that a set of plastic treads and accompanying sprockets and idlers is not cheap; whereas the Tamiya tracks retail mail order for under $12, expect to pay upwards of $50 to $60 for the plastic variety. However, once fitted with these plastic tracks, your Tunebot will be able to traverse a wider variety of surfaces, including fairly thick carpet, and yes, even super-polished floors.

Chapter 11

Going Places with the Telebot

The traditional robot is a mechanical contraption that operates under its own volition. The robot's maker uploads a program to its brain, disconnects the cable, and lets the beast roam free.

And yet that's only one type of robot. Another is the remotely operated kind, where a human guides the actions of the bot from afar. A link, wired or wireless, keeps maker and robot in constant communication. If the bot is in another room—or even another building, city, state, or country—a video camera beams back pictures so that what the machine sees, its human operator can see as well.

This is *telerobotics*: *tele* for distant, and *robotics* for, well, you know that part already. A telerobot is remotely controlled, though it may also contain autonomous functions that operate without human intervention. Some telerobots are commanded within the space of a living room, and some millions of miles away on alien planets.

In this chapter you'll build the Telebot, a convergence of ordinary robot with remote control abilities. The Telebot is based on the Teachbot DC, described in Chapter 9. This robot in turn uses many of the same parts as the Teachbot Servo platform that was detailed in Part II of this book.

Figure 11-1 shows the completed Telebot with handheld remote control and optional video broadcaster on top.

Be sure to visit the ARB Support Site (see Appendix A) for additional code examples and variations. All sketches from this book are available for ready download.

Figure 11-1 The Telebot is a Teachbot DC with a wire radio link and a wireless video camera capable of superimposing text over the picture. Beside the Telebot is the wireless remote control, which uses its own Arduino.

Using Radio Waves to Control a Robot

There are numerous methods to control a robot. For this chapter I'll demonstrate just one way, using a radio link. But for background, here are a few other common schemes:

- **Wired link** Connect a control panel consisting of buttons and switches directly to the robot's motors. The finished result is more a motorized toy than a real robot, because you directly influence every action of the bot—it has no independent control.
- **Infrared remote control** A universal TV remote can be used to send coded signals to a robot. A receiver mounted on the bot receives the signal, and a microcontroller commands the robot based on which buttons on the remote were pressed. Because the robot contains a microcontroller, it is capable of independent control in addition to human-influenced control. For example, you might press the button to have the robot go forward, but sensors on the bot will detect collision with something. The robot's own programming can reverse direction, without requiring any corrective action from you.
- **Line following** In this type of control a predefined line, dot, or other shape indicates where the robot is to go. Lines are usually drawn on paper, but it's also possible to interactively draw the lines by using such techniques as a laser pointer, or even a big flat-screen TV resting on its back.
- **Sound, light, or other sense** These are all *indirect* control methods, where the robot is influenced by some sensory condition you manage. For example, you might build a robot that has a light detector that can be "steered" using a flashlight or handheld laser pointer. A *musicbot* might be able to differentiate tones played on a piano or guitar.

For the Teachbot DC turned Telebot, you'll learn how to use the ZigBee standard of compact and affordable two-way radios. These radios operate in the high-frequency microwave band—but don't worry, they won't cook any part of you!—and work indoors or out.

GETTING STARTED WITH ZIGBEE

ZigBee is a wireless data standard. Two ZigBee radios form a link for sending and receiving serial data over the air. The standard ZigBee is based on the IEEE 802.15.4 specification, intended for low-speed, low-power wireless data. ZigBee is only one of the technologies that uses the 802.15.4 specification—MiFi and WirelessHART are a couple others—but it's the one that's been most embraced by the robotics community.

The idea behind ZigBee is keeping costs low, so that wireless communications can be affordably built into consumer products such as light switches and home theater receivers. As a result, ZigBee data radios are among the least expensive of them all. You can purchase a pair of them for under $50, depending on the feature set.

ZigBee operates at the 2.4-GHz (gigahertz) band, which is shared by a number of other wireless systems, including home Wi-Fi, cordless phones, and security video systems. Each ZigBee radio has a Data In pin (transmit) and a Data Out pin (receive). These two pins are referred to as DIN and DOUT, respectively.

Though the typical ZigBee radio module has almost two dozen connection pins, only power, ground, and one signal wire are needed to establish an over-the-air serial link.

Some ZigBee radios operate at 914 MHz (megahertz) and 915 MHz, depending on the international locale. These frequencies support much lower data rates than ZigBee radios that run at 2.4 GHz.

ZigBee radios are low-power; the entry-level modules output just 1 milliwatt of radio frequency (RF) signal. Depending on the module, frequency, and environmental factors, range is from about 30 to 250 feet. Because the RF output of ZigBee is low, it uses only a little current to operate. This helps to save on battery power, a great feature on any mobile robot.

However, if you do need to broadcast over a greater distance, you can opt for a higher output radio. The PRO modules with 63 milliwatts (mW) provide one to two miles of range, depending on terrain and data speed—you get greater range with slower speeds.

A self-contained module that conforms to the ZigBee standard, and complete with integrated antenna and mounted on a carrier board, is shown in Figure 11-2.

UNDERSTANDING VARIATIONS IN ZIGBEE STANDARDS

Complicating matters somewhat is that ZigBee encompasses both a basic and extended set. Actually there are more variations than these two, but many of the others are proprietary or special-purpose. If you're interested in learning more, check out the Wikipedia page on ZigBee for more information. Radios of different sets aren't compatible with one another, so be sure to get two modules that match.

- Basic ZigBee is often referred to simply as 802.15.4 (the IEEE standard that defines it), Series 1, or S1.
- The "extended" ZigBee is better known as ZB. It uses a different and more complicated communications protocol, among other differences.

Figure 11-2 An XBee IEEE 802.15.4 (Series 1) radio, on a Parallax carrier board. The board provides regulated 3.3V to the XBee.

As a side benefit of its low-cost nature, many ZigBee modules designed for hobby and experimental use also include a number of handy data acquisition features, highly useful for robotics. For example, the module might have one or more of its own analog-to-digital (ADC) inputs, as well as multiple digital input/output pins. I won't be talking about these features here, but you should know they're available should you need them.

USING XBEE SERIES 1 RADIOS

The Telebot uses a pair of low-cost XBee Series 1 modules. XBee is a popular brand of ZigBee RF radios that support the 802.15.4 protocol. The form factor—the physical layout—of the XBee radios is widely supported with various adapters that allow you to plug them directly into solderless breadboards. As the transmission range need not be extensive, the low-power 1 milliwatt (1mW) modules are more than adequate.

XBee modules use connector pins that are set 2mm apart. In order to plug the XBee module into a solderless breadboard you must first connect it to a carrier or adapter. These carriers come in various styles, with a variety of features.

Series 1 modules are simple to use. If you don't mind the default 9600 baud communications speed, they provide out-of-the-box factory settings for quick and easy setup. While you can always modify the factory settings for such things as communications speed, channel, and ID number (so more than two XBees can link at the same time), none of this is required if you're only needing to establish a basic wireless serial link between two nodes (points).

Three Arduinos, One Telebot

Telebot uses several Arduino controllers:

- The main Arduino (specifically, an Adafruit Boarduino) on the Telebot provides the primary functions of the robot, such as motor control.
- A remote control sends wireless commands to the Telebot. The remote uses any of several sensors to provide actuation. Options include a five-position switch and a tilt-compensated compass. A separate Arduino is used to process these sensors. Both the Telebot and the remote control are outfitted with XBee radios for transmitting the commands from the remote.

An optional data collection platform, using a separate Arduino, combines real-time video with the display of environmental information, such as gas levels, or temperature and humidity. The video and text display are transmitted wirelessly to a monitor.

Setting Up the Telebot Remote

The remote for the Telebot consists of an Arduino Uno development board, prototyping shield with mini solderless breadboard, battery, Series 1 XBee radio, and at least one of the following:

Five-position switch   The most straightforward control is with a mechanical switch. The five-position switch is literally five switches in one, arranged to provide up/down/left/right navigation. (The fifth switch is a center pushbutton.) Each of the five switches has its own pull-up resistor, and is interfaced to the Arduino with only a simple wire connection.

Two-axis accelerometer   The accelerometer allows you to tilt the remote forward and back and left and right to operate the Telebot. You need a two-axis module so that the sensor reacts to tilt on both the X and Y planes.

Tilt-compensated compass Using a compass lets you steer the Telebot by rotating the remote within the Earth's magnetic poles, or by tilting. You need a matching compass on the Telebot so that when the remote points east, so will the Telebot. Tilt compensation is used to correct for errors in the compass reading when the remote is not held level. But the compass also contains its own two-axis accelerometer, providing the same functionality as that described earlier. For my prototype Telebot remote I used a Devantech CMPS10 module.

MOUNTING THE ARDUINO AND PROTOSHIELD

Begin construction of the Telebot remote by cutting and drilling the base plate as shown in Figure 11-3. Mount the Arduino to the plate using a set of four 1/2" long nylon standoffs and 4-40 machine screws. Use 4-40 × 1/2" flat head screws on the underside of the plate, and 4-40 × 3/8" screws to mount the Arduino board to the standoffs. If using metal screws add a plastic washer to prevent any possible shorts.

Attach the protoshield to the Arduino. If the shield does not already have a mini breadboard on it, attach one using double-sided foam tape (most breadboards have the tape already applied).

CONNECTING THE XBEE RADIO

All three of the control sensors are linked, by way of the Arduino, to a Series 1 XBee radio. The wiring diagram is shown in Figure 11-4. The radio is mounted on a 22-pin carrier. Remember that XBees are designed for 3.3V operation. I'm using a carrier made for either 3.3V or 5V operation. The carrier has its own built-in 3.3V regulator, so you can use the Arduino's 5V supply.

Why not just use the Arduino's 3.3V power supply? It's a matter of power consumption. The 3.3V regulator on the Arduino Uno and similar boards is limited to providing 50 mA of current. That's about what the XBee uses when transmitting—that's too close for comfort.

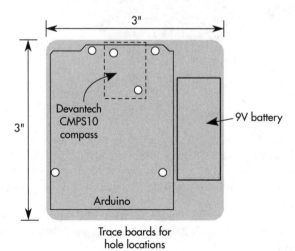

Figure 11-3 Layout guide for the remote baseplate. Use the Arduino and (optional) Devantech CMPS10 compass for marking holes for drilling.

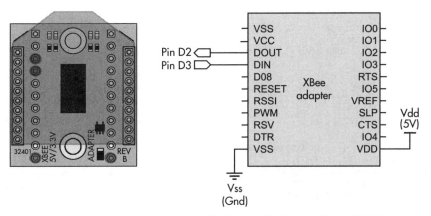

Figure 11-4 Connection diagram for the XBee radio (mounted on a 3.3/5V carrier board from Parallax) to remote Arduino.

> Though the XBee works on 3.3V, the input and output pins on the module are 5V tolerant. That means they can be directly connected to a 5V device, like the Arduino Uno, without the need for current-limiting resistors or level-shifting electronics.

USING THE FIVE-POSITION SWITCH

The wiring diagram for the five-position switch is shown in Figure 11-5. The switch comes on an eight-pin breakout board that's a bit oversized but will still fit onto the breadboard with the XBee module. Programming code is provided in *Telebot_Remote_SwitchOnly*. The sketch sends out a single byte character, indicating the direction of the switch press: *u* for up, *d* for down, and so forth.

Telebot_Remote_SwitchOnly

```
#include <SoftwareSerial.h>

// Pin connections for switch (includes power and ground)
const int pwrPin = 7;
const int gndPin = 4;
const int upBttn = 5;
const int crBttn = 6;
const int rtBttn = 8;
const int dnBttn = 9;
const int ltBttn = 10;
```

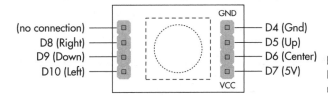

Figure 11-5 Wiring for the Parallax five-position switch to the remote Arduino

```
// Pin connections for XBee radio
const int xb_rx = 2;
const int xb_tx = 3;
SoftwareSerial Xbee(xb_rx, xb_tx);

int keyDelay = 300;          // Limit keypress to one per 300ms
long previousMillis = 0;     // Stores last time switch updated

void setup() {
  Xbee.begin(9600);

  // Set power to 5-position switch
  pinMode(pwrPin, OUTPUT);
  digitalWrite(pwrPin, HIGH);
  pinMode(gndPin, OUTPUT);
  digitalWrite(gndPin, LOW);
}

void loop() {
  unsigned long currentMillis = millis();

  // Prevents bounce
  if(currentMillis - previousMillis > keyDelay) {
    previousMillis = currentMillis;
    if(digitalRead(ltBttn) == LOW)
      Xbee.print("l");

    if(digitalRead(rtBttn)== LOW)
      Xbee.print("r");

    if(digitalRead(upBttn) == LOW)
      Xbee.print("u");

    if(digitalRead(dnBttn) == LOW)
      Xbee.print("d");

    if(digitalRead(crBttn) == LOW)
      Xbee.print("c");

  }
}
```

USING THE TWO-AXIS ACCELEROMETER/COMPASS

See Figure 11-6 for the wiring diagram. I'm using the integral accelerometer in the Devantech CMPS10 tilt-compensated compass. This device returns three pieces of information:

- Compass bearing, a value from 0 to 359. (Actually, the bearing is returned in two bytes—a low byte and a high byte—and these two bytes are combined to provide the bearing value. The combining is done in the receiver code, presented later.)
- Pitch, a positive and negative value indicating front and back tilt (0 means flat, or no tilt).
- Roll, a positive and negative value indicating side-to-side tilt (0 = flat).

The pitch and roll values are *signed* bytes, meaning that the number is positive if the byte value is 0 to 127, and negative if it's 255 to 128. Signed bytes are supported on the Arduino with the *char* data type.

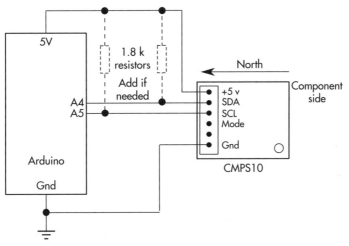

Figure 11-6 Wiring for the Devantech CMPS10 tilt-compensated compass (combines magnetometer plus two-axis accelerometer) to the remote Arduino

The CMPS10 compass is connected to the Arduino via I2C. Note the 1.8 kΩ pull-up resistors. These are technically called for, in accordance with I2C standards, but many devices will connect to the Arduino without them. Add them if communications don't work, or are sporadic.

 Telebot_Remote_CompassOnly

```
#include <SoftwareSerial.h>
#include <Wire.h>                          // Use I2C library
#define ADDRESS 0x60                       // Address of CMPS10

const int xb_rx = 2;
const int xb_tx = 3;
SoftwareSerial Xbee(xb_rx, xb_tx);

void setup() {
  Xbee.begin(9600);
  Wire.begin();                            // Connect to I2C
}

void loop(){
  byte bytehigh, bytelow;                  // bearing high, low bytes
  char pitch, roll;                        // Pitch, roll
  int bearing;                             // Full bearing

  Wire.beginTransmission(ADDRESS);         // Begin comm with CMPS10
  Wire.write(2);                           // Start read
  Wire.endTransmission();
  Wire.requestFrom(ADDRESS, 4);            // Request 4 bytes
  while(Wire.available() < 4);             // Wait for bytes to arrive
  bytehigh = Wire.read();                  // Store values
```

```
bytelow = Wire.read();
pitch = Wire.read();
roll = Wire.read();

Xbee.write(byte(0));
Xbee.write(bytehigh);
Xbee.write(bytelow);
Xbee.write(pitch);
Xbee.write(roll);

delay(500);
}
```

USING MULTIPLE SENSORS

The mini breadboard is too small to support more than one sensor along with the XBee radio. On my prototype I inserted both the XBee and five-position switch on the breadboard (see Figure 11-7), and mounted the compass on the base plate itself. Only one sensor at a time provides control feedback; putting both on the remote is merely a convenience, and you may choose one or the other.

If you'd like to switch back and forth between using the five-position switch and accelerometer/compass, use a three-pin jumper to select the mode. In one position the remote is set to use the five-position switch; in the other position the remote used the accelerometer

Figure 11-7 The completed remote control, with Arduino, protoshield with mini breadboard, 9-volt battery, XBee radio, and five-position switch. The CMPS10 compass is mounted on the baseplate in front, out of view.

or compass. As there's a dearth of power (+V and Gnd) pins available, the mode jumper uses adjacent pins for the reference voltage.

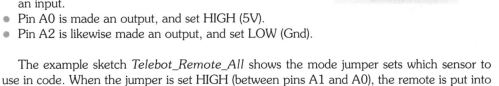

- Analog pin A1 is the mode setting pin, and is left as an input.
- Pin A0 is made an output, and set HIGH (5V).
- Pin A2 is likewise made an output, and set LOW (Gnd).

The example sketch *Telebot_Remote_All* shows the mode jumper sets which sensor to use in code. When the jumper is set HIGH (between pins A1 and A0), the remote is put into *button* mode. Conversely, when the jumper is set LOW (between pins A1 and A2), the remote is put into *compass* mode. Remember: The CMPS10 module contains both a magnetometer for compass bearing and accelerometer for tilt.

Telebot_Remote_All

```
#include <SoftwareSerial.h>
#include <Wire.h>                    // Use I2C library
#define ADDRESS 0x60                 // Address of CMPS10

// Switch pinouts
const int pwrPin = 7;
const int gndPin = 4;
const int upBttn = 5;
const int crBttn = 6;
const int rtBttn = 8;
const int dnBttn = 9;
const int ltBttn = 10;

// Mode pinouts
const int mode_set = A1;
const int mode_high = A0;
const int mode_low = A2;

// XBee pinouts
const int xb_rx = 2;
const int xb_tx = 3;
SoftwareSerial Xbee(xb_rx, xb_tx);

int keyDelay = 300;
long previousMillis = 0;

void setup() {
  Serial.begin(9600);
  Xbee.begin(9600);
  Wire.begin();

  // Set power to 5-position switch
  pinMode(pwrPin, OUTPUT);
  digitalWrite(pwrPin, HIGH);
  pinMode(gndPin, OUTPUT);
  digitalWrite(gndPin, LOW);

  // Set operating mode (switch or compass)
  pinMode(mode_high, OUTPUT);
  digitalWrite(mode_high, HIGH);
```

```
    pinMode(mode_low, OUTPUT);
    digitalWrite(mode_low, LOW);
}

void loop() {
  if (digitalRead(mode_set) == HIGH)
    buttons();
  else
    compass();
}

void buttons() {
  unsigned long currentMillis = millis();
  if(currentMillis - previousMillis > keyDelay) {
    previousMillis = currentMillis;

    if(digitalRead(ltBttn) == LOW) {
      Serial.println("left");
      Xbee.print("l");
    }

    if(digitalRead(rtBttn)== LOW) {
      Serial.println("right");
      Xbee.print("r");
    }

    if(digitalRead(upBttn) == LOW) {
      Serial.println("up");
      Xbee.print("u");
    }

    if(digitalRead(dnBttn) == LOW) {
      Serial.println("down");
      Xbee.print("d");
    }

    if(digitalRead(crBttn) == LOW) {
      Serial.println("center");
      Xbee.print("c");
    }
  }
}

void compass() {
  byte bytehigh, bytelow;
  char pitch, roll;
  int bearing;
  Wire.beginTransmission(ADDRESS);
  Wire.write(2);
  Wire.endTransmission();
  Wire.requestFrom(ADDRESS, 4);
  while(Wire.available() < 4);
  bytehigh = Wire.read();
  bytelow = Wire.read();
  pitch = Wire.read();
  roll = Wire.read();

  //Send through XBee
  Xbee.write(byte(0));
  Xbee.write(bytehigh);
  Xbee.write(bytelow);
```

```
        Xbee.write(pitch);
        Xbee.write(roll);

        bearing = ((bytehigh<<8) + bytelow) / 10;   // Calc bearing
        print_data(bearing, pitch, roll);           // Print data

        delay(500);
}

// Print data to Serial Monitor window (debugging)
void print_data(int bearing, int pitch, int roll) {
    Serial.print(bearing, DEC);
    Serial.print(":");
    Serial.print(pitch, DEC);
    Serial.print(":");
    Serial.println(roll, DEC);
}
```

The *Telebot_Remote_All* sketch also includes a data visualizer, so you can see the data being sent to the XBee. Open the Serial Monitor to verify operation. Note the line in the sketch

```
        bearing = ((bytehigh<<8) + bytelow) / 10;
```

which combines the high and low compass bytes to produce the single 0 to 359 bearing value. The same line of code is used in the receiving sketch on the Telebot robot.

Adding an XBee Receiver to the Telebot

To turn the Teachbot DC into the Telebot you need only to add an XBee module to receive the signals from the remote control unit. To save space on the breadboard atop the robot I've elected to use a SIP-style XBee carrier, shown in Figure 11-8. The carrier board lacks mounting holes (at least the kind I'm using), but it can be attached using a cable tie threaded through two holes.

Electrical connection requires merely four wires (see Figure 11-9), two for power and one each for data receive and transmit.

As noted earlier in the chapter, when using two Series 1 XBee radios, you can keep the default factory settings, and the two radios in your link will automatically know how to communicate with one another. Transfer speed is 9600 baud, which is plenty fast enough for the kind of simple data used for the Telebot. A speed of 9600 baud is over 900 bytes per second, and the most data sent from the remote is 5 bytes at a time.

COMMANDING THE TELEBOT WITH THE FIVE-POSITION SWITCH

The *Telebot_Receiver_Switch* sketch shows how to interface the XBee receiver to the serial motor controller used on the Telebot. Note the use of the *SoftwareSerial* object library, which comes with the Arduino IDE (version 1.0), and the *CompactQik2s9v1* library, available from Pololu.com—they make the Qik2s9v1 serial motor controller used on the Telebot. The library is also provided on the ARB Support Site (see Appendix A). If fetching the library from the Pololu Web site be sure it's the version for the Arduino 1.0 IDE software.

Figure 11-8 The XBee radio on the Telebot is mounted on a SIP adapter and secured to the send base using an ordinary cable wrap.

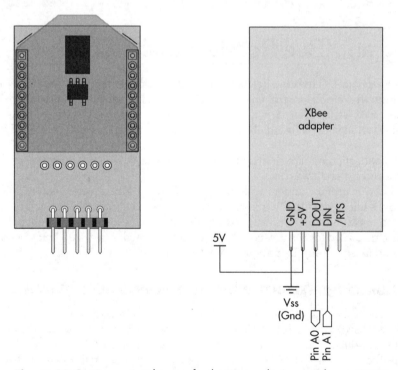

Figure 11-9 Connection diagram for the XBee radio (mounted on a 5V SIP carrier board from Parallax) to Telebot Arduino

To use this sketch be sure to first place the *CompactQik2s9v1* library files in the Arduino sketch *libraries* folder.

> The *SoftwareSerial* library provides serial communications using any Arduino pin. It's useful for when you want to preserve the Arduino's hardware serial port (on pins D0 and D1), and you wish to string multiple serial-based hardware off your Arduino. This sketch, like all sketches in this book, is intended for Arduino 1.0 or later.

Telebot_Receiver_Switch

```
#include <SoftwareSerial.h>
#include <CompactQik2s9v1.h>

#define xbeeRx    A0          // DOUT to A0
#define xbeeTx    A1          // DIN to A1
#define rxPin     5           // Qik motor Rx
#define txPin     6           // Qik motor Tx
#define rstPin    7           // Qik motor reset

byte topSpeed = 96;           // 0 to 127

SoftwareSerial motorSerial (rxPin, txPin);
CompactQik2s9v1 motor = CompactQik2s9v1(&motorSerial, rstPin);
SoftwareSerial Xbee (xbeeRx, xbeeTx);      //After motor serial

boolean motorsOn = false;

void setup() {
  Serial.begin(9600);
  motorSerial.begin(9600);
  motor.begin();
  motor.stopBothMotors();
  Xbee.begin(9600);           // Place last
  delay(200);
  pinMode(13, OUTPUT);        // Show ready status
  digitalWrite(13, HIGH);
}

void loop() {
  readXbee();
  delay(50);
}

void readXbee() {
   if(Xbee.available()) {
     char val = Xbee.read();
     Serial.println(val);
     controlMotor(val);
   }
}

void controlMotor(char val) {
  switch (val) {
    case ('c'):
      motorsStop();
      break;
    case ('u'):
      motorsForward(topSpeed);
      break;
```

```
      case ('d'):
        motorsReverse(topSpeed);
        break;
      case ('r'):
        motorsTurnRight(topSpeed);
        break;
      case ('l'):
        motorsTurnLeft(topSpeed);
        break;
  }
}

//  motor0 = left, motor1 = right
void motorsStop() {
  motor.motor0Forward(0);
  motor.motor1Forward(0);
}

void motorsForward(byte motorSpeed) {
  motor.motor0Forward(motorSpeed);
  motor.motor1Forward(motorSpeed);
}

void motorsReverse(byte motorSpeed) {
  motor.motor0Reverse(motorSpeed);
  motor.motor1Reverse(motorSpeed);
}

void motorsSpinRight(byte motorSpeed) {
  motor.motor0Forward(motorSpeed);
  motor.motor1Reverse(motorSpeed);
}

void motorsSpinLeft(byte motorSpeed) {
  motor.motor0Reverse(motorSpeed);
  motor.motor1Forward(motorSpeed);
}

void motorsTurnRight(byte motorSpeed) {
  motor.motor0Forward(motorSpeed);
  motor.motor1Forward(0);
}

void motorsTurnLeft(byte motorSpeed) {
  motor.motor0Forward(0);
  motor.motor1Forward(motorSpeed);
}
```

Because of the way software-based serial communications work, using multiple serial objects at the same time can cause conflicts within the Arduino hardware. This may limit how many of these objects are used in your sketches. The Telebot uses two *SoftwareSerial* objects: one for the XBee radio, and one for the serial motor controller.

During development of this project I noticed that the order in which the two objects appear in the sketch influenced how well the two got along. The XBee serial object would not work if it was defined and started prior to the motor serial object.

When these types of conflicts cannot be resolved by simply relocating code, you'll need to either select a simpler software serial library, or use the Arduino's built-in hardware serial port.

Points of interest in the sketch:

- The *setup()* routine starts three serial objects (one for the XBee, two for the motor controller).
- The *readXBee* function continuously reads the XBee receiver, and if there's data, processes it using the *controlMotor* routine.
- Then in the *controlMotor* routine the bytes received from the XBee are compared against the five expected single-letter values. Each value has a corresponding motor routine. For example

```
case ('u'):
  motorsForward(topSpeed);
```

runs both motors forward when receiving the *u* (Up) signal.

COMMANDING THE TELEBOT WITH COMPASS TILT

The *Telebot_Tilt_Receive* sketch demonstrates receiving compass bearing and tilt data from the remote. The first half of this sketch is identical to the previous one so the code is omitted to save page space. New code is provided for the *readXbee, controlMotor,* and various motor functions.

To operate the Telebot, program the remote for tilt mode, then

- Slowly pitch it forward to move the robot forward. The more you tilt the remote, the faster the motors will go.
- The same happens when you pitch the remote backward. The robot goes into reverse.
- Level out the remote to stop the robot.
- Tilt the remote to the right or left to make a turn. To keep the code simple the Telebot's speed is still maintained by the pitch angle. To turn you must also tilt the remote from side to side. Note that for simplicity (and to make it easier to drive the Telebot), there is no turning in reverse.

The motors are not activated unless the pitch value is greater than 5 or –5. I'm using a simple scaling formula to turn the pitch data into motor speed, which is from 0 to 127. You can experiment with this scale—with the line *min(abs(pitch)*3, 127)*—to use a greater or lower scale. The Arduino *min* (for *min*imum) function is there to prevent the value from exceeding 127. The Arduino *abs* (for *abs*olute) function ensures the value is always positive.

Instead of scaling the pitch value, you can map it, using the Arduino *map* function. This function translates the pitch values—5 to 50—to speeds 0 to 127. The Arduino *constrain* function is added to ensure the speed never goes above 127:

```
topSpeed = constrain(map(abs(pitch), 5, 50, 0, 127), 0, 127);
```

Furthermore, the code is simplified so that any roll value greater than 20 or –20 places the robot in a turn.

Telebot_Tilt_Receive

```
// Code same as Telebot_Receiver_Switch

void readXbee() {
  byte bytehigh, bytelow;    // High and low bytes of bearing
  char pitch, roll;          // Pitch and roll (signed values)
  int bearing;               // Full bearing

  if(Xbee.available() == 5) {
    Xbee.read();                      // Throw away first
    bytehigh = Xbee.read();           // Store other values
    bytelow = Xbee.read();
    pitch = Xbee.read();
    roll = Xbee.read();
    controlMotor(pitch, roll);        // Control motors
  }
}

void controlMotor(char pitch, char roll) {
  topSpeed = min(abs(pitch)*3, 127);
  if(pitch > 5) {
    motorsReverse(topSpeed, roll);
  } else if(pitch < -5) {
    motorsForward(topSpeed, roll);
  } else {
    motorsStop();
  }
}

//  motor0 = left, motor1 = right
void motorsStop() {
  motor.motor0Forward(0);
  motor.motor1Forward(0);
}

void motorsForward(byte motorSpeed, char turn) {
  if(turn > 20) {
    motor.motor0Forward(motorSpeed);
    motor.motor1Forward(0);
  } else if(turn < -20) {
    motor.motor0Forward(0);
    motor.motor1Forward(motorSpeed);
  } else {
    motor.motor0Forward(motorSpeed);
    motor.motor1Forward(motorSpeed);
  }
}

void motorsReverse(byte motorSpeed, char turn) {
  motor.motor0Reverse(motorSpeed);
  motor.motor1Reverse(motorSpeed);
}
```

The *Telebot_Tilt_Receive* sketch lacks any error detection or resynchronization code for ensuring that the four bytes from the remote are always received in the proper order. Under normal operating conditions error detection and resynchronization should not be necessary. The sketch checks when it's received five bytes (which always start with 0). When it has, the data is immediately processed.

One way to ensure proper synchronization between remote and Telebot is to determine if there's more than five bytes waiting in the Serial buffer:

```
Xbee.available() > 5
```

That means that somehow a few extra bytes have arrived. Resync by flushing the Serial buffer (use Serial.flush), and waiting a minimum of 50 milliseconds for all data transmission to stop. Any data that comes in after such a delay indicates it's a new group of bytes. The minimum wait helps avoid capturing bytes in midstream of a five-byte packet from the remote.

On the Web: Commanding the Telebot with Compass Bearings

Lastly, but certainly not least, you can control the direction of the Telebot using the bearing value from the CMPS10 compass. You need to add a second compass to the Telebot itself, so that the robot knows which direction it's pointing.

In operation, the Telebot examines the bearing from the remote, compares it against the value of its own internal compass, and spins the robot until the two are (more or less) the same. You still control the speed of the robot by pitching the remote forward or back.

With space in this chapter growing short, the compass-bearing plans and sketch are placed on the ARB Support Site—see Appendix A for details. There you'll also find an enhanced version of the *Telebot_Tilt_Receive* sketch in the preceding section. The upgraded sketch provides added debugging features so you can check the values being received by the Telebot's XBee receiver.

Broadcasting Real-Time Video with the Telebot

Let's go even further with the Telebot: In addition to controlling the robot remotely, how about having the Telebot send back video pictures of the landscape it sees. Wireless security cameras are relatively low cost—some under $40, including both camera/transmitter and receiver—and easy to use on a mobile robot. The camera is operated using a battery or other convenient power supply.

Figure 11-10 shows a small wireless camera mounted on a third deck of the Telebot, along with the receiver for the camera. The second and third decks are separated by long standoffs. The camera is operated by its own 9-volt battery; the receiver has a plug-in power adapter. The video output of the receiver connects to any compatible monitor. As I'm using an NTSC camera and receiver, any NTSC-compatible monitor will do. A small 7" LCD monitor is ideal.

The standard frequency band used by wireless video cameras is 2.4 GHz, which so happens is the same band used by your XBee radios. There is a chance that the video camera will interfere with the XBee, and vice versa. However, this problem is mitigated by using a multichannel wireless camera, where you can select a different RF frequency within the 2.4 GHz band.

You can tell if your camera and XBee are lousing up one other by monitoring the video image when you place the XBee remote next to the camera. If there's a problem you'll see wavy lines and other interference patterns.

If your wireless camera lacks a channel-changing feature, or selecting none of its channels removes the interference, you can try altering the XBee's channel; there are 16 to choose from, and each channel occupies its own 5 MHz space. The channel on both the transmitter and receiver XBee must be tuned to the same channel.

Refer to the documentation on the XBee radios on how to set channels. It can be done directly in the Arduino sketch (using something called *Command Mode*), though most people prefer to use the separately available X-TCU program, which provides a handy graphical interface. You need an XBee carrier board with USB in order to communicate between your PC and the radio. The X-TCU program is available for download from *digi.com,* makers of the XBee.

Figure 11-10 Hacked backup camera, showing the separate transmitter and camera. It's ugly, but it works. (I don't advise you to try the same; it seems backup cameras use quite a bit of weatherproofing, making them undesirable hacking candidates.)

SUPERIMPOSING TEXT OVER VIDEO

Not only can your bot beam back video, you can superimpose text and other images onto the video, a kind of *Terminator*-style view of the world around it. The technique involves passing the video through an *overlay* device, an electronic circuit that's capable of locking onto the incoming video signal and superimposing text over it.

While you can use an Arduino to overlay text onto video, the process involves added circuitry, and the heavy demands of video processing leave the Arduino little time to do anything else. So instead, I decided to use a Propeller Backpack (available from *parallax.com*). The Backpack is a video generating and processing add-on that connects to most any microcontroller. You need only attach a video source to the input, route the output to a wireless transmitter or monitor, and use simple serial commands to place text anywhere over the image.

There are two ways to connect the Backpack to the video circuitry:

- **Between the camera and transmitter**   This works best if your wireless system has a separate camera and transmitter, and both are reasonably high quality. If your wireless camera is all-in-one, you need to tap into the video line to make the connection to the Backpack. Not impossible, but depending on the camera, such electronic "surgery" can get tricky. You might even wreck the camera attempting it, so be careful.
- **Between the transmitter and the monitor**   In order for this to work you need to use the XBee radio on the Telebot to send back data to your base station. You can use the Arduino on the remote control to receive the data, but you'll need to add a bigger breadboard for the Backpack, or else mount the Backpack elsewhere. The Backpack and Arduino are connected via two short wires. (One idea: Put the Backpack under the Arduino. Increase the length of the standoffs if you need more room.)

For the Telebot prototype I elected to put the Backpack between the camera and transmitter. I found a wireless car backup camera—the kind you mount above the rear license plate—and hacked it to get to the video connections inside. This proved to be quite difficult, as the camera was weatherproofed and greatly resisted deconstruction.

After tapping through to separate the video wire (it's nearly always white or yellow, by the way), I mounted the camera with lens, and the transmitter, on a piece of 1/8" plastic, then attached that using brackets to the third deck of the Telebot. The video input from the camera connects to the Backpack, which then connects to the transmitter, as shown in Figure 11-11.

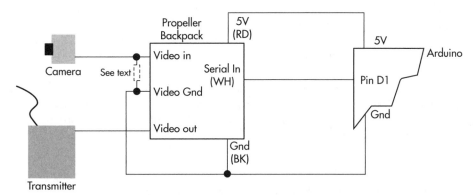

Figure 11-11 Wiring diagram for the Parallax Propeller Backpack video overlay module

Plug

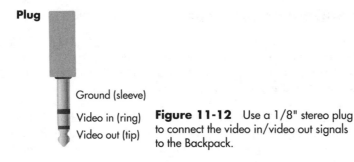

Ground (sleeve)

Video in (ring)

Video out (tip)

Figure 11-12 Use a 1/8" stereo plug to connect the video in/video out signals to the Backpack.

You'll need to make a connector for piping the video through the Backpack. Use a 1/8" stereo plug, and wire the video in, video out, and ground connections as shown in Figure 11-12. Keep wire lengths short; the shorter the better to reduce signal noise and interference. If you have it, three-conductor shielded cable (two inner conductors, one outer shield) will provide better immunity to noise. The outer shield is connected to ground.

Not all cameras output perfect video. The Propeller Backpack requires a fairly standard video signal, or else it may not be able to properly overlay text. If the level of the video signal is too high, the text may not appear, or may appear overly dark or light.

If that happens, try playing a small 1/8 watt resistor across the camera input, as shown in Figure 11-11. Experiment with the value of the resistor, but a good starting point is 150 Ω.

WRITING TEXT TO THE BACKPACK

Before you can use the Propeller you must first reprogram it to work in overlay mode (it comes preprogrammed for displaying text over a black or colored background). In order to reprogram the Backpack you'll need a Prop Plug, also from Parallax, to connect your PC to the Propeller via USB. Fetch and install the Parallax Propeller Tool software from the *parallax.com* site, and locate the overlay programming code—it's provided as a link on the Backpack page. Use this code in the Propeller Tool software to upload the new programming to the Backpack.

Once reprogrammed, you can mount the Backpack to the third deck of the Telebot. You need only use one I/O line for basic communications with the Backpack. I elected to use the Arduino's hardware serial port to connect to the Backpack. Only the D1 transmit (Tx) pin is connected; the pin D0 receive (Rx) pin is left unattached.

You can keep the Backpack connected to the Arduino while uploading sketches to it. The Backpack will also receive the programming data, but will be unable to make much sense of it. In some cases, the intercepted data may cause the Backpack to lock up. If this occurs, after upload is complete momentarily disconnect power, and let the Arduino and Backpack reset.

To write text to the Backpack you first set up what the Backpack documentation refers to as *displays* and *windows*. Once these have been defined, you can write text into the display. The sketch *Telebot_Backpack* shows a *Hello World* example of preparing the Backpack, then displaying a line of text. To make it easier to use the Backpack I've created several functions that encapsulate common tasks.

Telebot_Backpack

```
const int backPack = 1;          // Pin D1 for Backpack serial

//Define constants for Backpack
#define APNDSP      0x18
#define CHGCLR      0x12
#define CLS         0x00
#define DEFWIN      0x10
#define HOME        0x01
#define LF          0x0a
#define MOVWIN      0x19
#define SHODSP      0x07
#define SHOWIN      0x1a
#define USEWIN      0x11
#define WDWRAP      0x15

#define DBL         0x80
#define SHO         0x40
#define TPT         0x01

void setup() {
  // Reset Backpack
  pinMode(backPack, OUTPUT);
  digitalWrite(backPack, LOW);
  delay(100);
  pinMode(backPack, INPUT);
  delay(2000);

  // End reset Backpack

  Serial.begin(9600);
  delay(1000);

  byte windowA = 1;
  byte displayNum = 2;
  defineWindow(windowA, 44, 2);    // Define window 44 x 2
  changeColor();
  appendDisplay(windowA, displayNum, 1, 1);
  Serial.println("This text displayed on the Backpack");
  delay(2000);                     // Wait 3 seconds
  showDisplay(0);                  // Hide text
}

// Empty loop; for this demo everything is done in setup
void loop() { }
```

```
// Backpack helper functions follow

// Define a window, given a window number,
//    number of columns, number of rows

void defineWindow(byte windowNum, byte cols, byte rows) {
  Serial.write(DEFWIN);
  Serial.write(windowNum);
  Serial.write(cols);
  Serial.write(rows);
}

// "Append" a window to a display, given a window number
//    display number, and x and y coordinates on the screen

void appendDisplay(byte windowNum,
byte displayNum, byte x, byte y) {
  Serial.write(APNDSP);                 // Append display to window
  Serial.write(displayNum);             // display
  Serial.write(SHO | windowNum);        // winno
  Serial.write(x);                      // x
  Serial.write(y);                      // y
  Serial.write(SHODSP);                 // Show display
  Serial.write(displayNum);
  Serial.write(WDWRAP);                 // Turn on word wrap
  Serial.write(1);
}

// Change color of text and background
void changeColor() {
  Serial.write((byte)CHGCLR);
  Serial.write(0x0f);                   // mask
  Serial.write(0x08);                   // trans
  Serial.write(0x08);                   // foreground
  Serial.write((byte)0x00);             // background
}

// Move the window to a new coordinate on the screen
void moveWindow(byte displayNum, byte x, byte y) {
  Serial.write((byte)MOVWIN);
  Serial.write(displayNum);
  Serial.write(x);
  Serial.write(y);
}

// Use (activate) a window
void useWindow(byte windowNum) {
  Serial.write((byte)USEWIN);
```

```
      Serial.write(windowNum);
}

// Show (or hide) a display (shows or hides the text)
void showDisplay(byte displayNum) {
  Serial.write((byte)SHODSP);
  Serial.write(displayNum);
}

// Clears all text
void cls() {
  Serial.write((byte)CLS);
}
```

Be sure to *thoroughly* read the Propeller Backpack Video Overlay Object documentation, especially the General Concepts section, which describes displays, windows, and slots. You'll especially want to study how the Backpack keeps track of the current window and display, and how commands subsequently sent affect the current settings.

And note that (as of this writing) the Backpack firmware does not support two side-by-side windows for text. You can stack windows vertically, but you can't place them beside one another.

In operation, the sketch begins by first resetting the Backpack. This is done by making pin D1 an OUTPUT, and bringing the pin low for 100 milliseconds. The pin is then made an INPUT, so that it can be used by the Arduino hardware serial port. Then, and only then, is the hardware Serial port set up to transmit data to the Backpack.

From here, the sketch sets up a *window* and a *display* for holding a line of text. The numeric values you assign here are arbitrary; I'm using 1 and 2, respectively. The functions *defineWindow, changeColor,* and *appendDisplay* functions are ones I've defined; the routines themselves are at the bottom of the sketch. The functions merely send out bytes that prepare the Backpack display. For example, the *defineWindow* function sends four bytes to the Backpack:

```
Serial.write(DEFWIN);        // Define window command (0x10)
Serial.write(windowNum);     // Window number (e.g. 1)
Serial.write(cols);          // Number of text columns
Serial.write(rows);          // Number of text rows
```

DISPLAYING ENVIRONMENT DATA

Any sensor that provides printable text can be used with the Telebot to provide environmental data. As a demonstration I'll show using an LPG sensor to detect levels of natural gas. The sensor is easily connected to the Arduino (see Figure 11-13), and returns an analog value indicating the concentration of natural and other gases.

The LPG sensor demonstrated here is for *educational use* only. Don't trust it as the sole noxious gas detector in your home. As natural gas is lighter than air, it will rise to the ceiling and concentrate there, away from a low-lying robot like the Telebot.

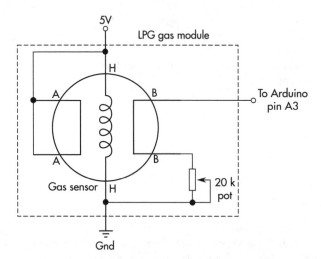

Figure 11-13 Wiring diagram for a typical LPG gas sensor; this one is an MQ5 from Seeedstudio. Convenient connection points are provided for 5V, ground, and output.

Other types of environmental sensors you might try include other types of gas sensors (alcohol, carbon dioxide, etc.), humidity, temperature, barometric pressure, compass heading, light level, and so on. The following code snippet assumes you've already set up the Backpack to display text. The Serial.print lines send data to the Backpack, as it's connected to the Arduino's hardware serial transmit pin.

```
void loop() {
  cls();                          // Clear old display
  int val = analogRead(A3);              // Read sensor
  Serial.print("Gas level: ");
  Serial.println(val, DEC);       // Print sensor value
  delay(1000);                    // Wait 1 second
}
```

I used an MQ5 LPG gas sensor module from Seeedstudio. This module is self-contained, providing connection points for 5V power, ground, and analog output. An on-board potentiometer allows you to adjust the sensitivity of the device. On initial use, dial the pot to its mid-point position, and adjust as needed.

Figure 11-14 shows the top of my prototype Telebot with the gas sensor attached, along with the Propeller Backpack, hacked wireless camera with transmitter board, and extra Arduino for processing everything.

Figure 11-14   Top view of the Telebot, showing the camera and transmitter, Backpack, gas sensor, breadboard, and Arduino. There's plenty of room left over for more sensors.

Telebot Enhancements

Before closing out this chapter I'd like to suggest a few neat features you might like to add to your Telebot. But don't just stop here—let your imagination soar!

- Add a speaker to the remote control unit to provide audible feedback. Use a piezo element for the loudest sound and least amount of current draw. Avoid using an 8 Ω dynamic speaker; it'll draw too much current. You can cut down on the current demand by installing a 100 to 150 ohm resistor in series between the speaker and Arduino, but then the sound may not be very loud.
- Mount a vibrator motor to the base plate, and interface it to the Arduino using a transistor driver (the ULN2003 driver chip is ideal for this; it contains seven separate drivers). Add some collision detectors to the Telebot, and when one is activated, use the XBee link to transmit an "I hit something!" signal. Briefly apply power to the motor when the remote gets the bump command.
- Use a Microsoft Kinect connected to your PC to transmit motion commands to the Telebot. You can do away with the Arduino on the remote, and instead connect the XBee transmitter (you'll need an XBee USB carrier) to your computer. The Microsoft Robotics Developer Studio (RDS) provides a programming platform for interfacing with and using the Kinect.

Chapter 12

Why Did It Have to Be Snakes?

Many of us love to hate snakes. Cold blooded, beady unflinching eyes, venomous fangs that can strike at lightning speed. And yet as creatures of the earth, they're among the most fascinating in the way they move. Snakes don't have legs; they travel by undulating their long, slender bodies over the ground, through water, or within trees or other foliage.

We've all seen robots that roll on wheels, ramble over treads, and walk with legs. But what about a machine that slithers and creeps, a robot that propels itself serpent-like over terra firma? In this chapter we'll look at one method of creating a multi-segmented robot that transports itself from here to there by making shimmying motions over the living room carpet.

I've designed the bot so that it isn't frightfully expensive—the most costly part is the 10 mini servo motors—and does not require specialized building tools. Still, this is not a project for beginners. There's a certain degree of precision required in constructing each of the snake's movable armatures. The programming is more complex than those in previous chapters, and involves considerable experimentation on your part. Tackle this one after you've got a couple of robots under your belt.

The snake design in this chapter is a prototype, with plenty of room to experiment and grow. The hardware and software I provide should get you started in constructing a snake-like robot that uses a rudimentary slithering action. If you decide to construct a snake robot of your own, I encourage you to share your designs and improvements with other Arduino robot enthusiasts. Sharing links are found on the ARB Support Site (see Appendix A).

Be sure to visit the ARB Support Site for code examples and variations. All sketches from this book are available for ready download.

How Snakes Move, Real and Robotic

The way a snake moves—called *locomotion*—depends much on the type of snake and the ground (or other medium) it's traveling over. There are nearly 3000 species of snakes, and some of them have developed unique and specialized movement. A robotic snake can emulate many of these locomotion principles, but not every means of movement is easy to replicate in a machine. The more sophisticated the locomotion, the more complex the mechanism.

For snakes, hardware copies nature by using multiple segments—the segments are linked one after the other like cars on a train. At each link is a motorized joint. Of the more basic robot snake designs—and the kind covered in this chapter—each joint has a single degree of motion (also called *degree of freedom,* or *DOF*). To mimic the slithering motion of snakes, the joint at each segment moves horizontally over the ground.

Of the various forms of snake locomotion, the following three are among the most common, though only the first method is typically found on robotic reptiles:

   Undulation is the familiar lateral (side-to-side) serpentine motion most commonly associated with snakes. The body of the snake makes full contact with the ground throughout its length. In mathematical terms, the undulation describes a sine wave, where the body of the snake is the crests and troughs of the wave. The exact shape of the wave depends on the type of snake, its size, its speed, and other factors.

   For a robot serpent, undulation provides the simplest means of locomotion, as it doesn't need complicated multi-axis segments. Movement requires only a single axis at each segment joint: one segment, one motor.

   Sidewinding combines a serpentine wave with up-and-down lifting. Locomotion occurs only at the points of the body that touch the ground. While a sidewinder robot looks supremely cool, the mechanics are significantly more complex than basic undulation. Each segment needs two joints, for both X and Y axes. So, given 10 segments, you need 20 motors, two for each segment joint.

   Concertina movement is where the snake binds up in a tight coil, then springs forward by straightening its body. Like undulation, concertina locomotion involves just one axis per segment, making it adaptable for low-end robotic reptiles. For added traction, the rear segment of the robot snake can incorporate a small "claw" that digs into the ground when the mechanism pushes forward.

Design Concept of the Snakebot

For the lack of a better name I'll call the snake robot detailed in this chapter Snakebot—it's descriptive even if the name isn't very creative. Figure 12-1 shows the completed Snakebot. Here's what it's all about.

SEGMENTS

Snakebot is made of 10 identical segments. Each segment (see Figure 12-2) consists of a base plate, mini servo, top plate, and associated fasteners to keep everything together:

- The base plate serves as a smooth surface for the bottom of the snake. But even with the smoothness, there's too much friction for easy movement, so wheels (they're actually

Figure 12-1 The completed Snakebot, showing its 10 articulated segments and distributed power and electronics

Figure 12-2 Each segment uses a mini servo set in a framework of a top and bottom plate. The plates are separated using a pair of 1" long 4-40 threaded standoffs.

common in many mechanical snakes) are added on either side of each segment. In my initial design I used small plastic wheels, but switched them out for larger rubber-treaded wheels, which provided better traction.

- The mini servos rotate the segments to provide the sinusoidal action for locomotion. I picked mini- rather than micro-sized servos for their greater turning power. Fully loaded with electronics and batteries, Snakebot weighs in at about two pounds.
- A top plate, made of a thinner material than the rest of the body, forms the upper mounting deck of each segment. On the plates go the various electronics and battery packs. A block diagram of the electronic subsystems is shown in Figure 12-3.

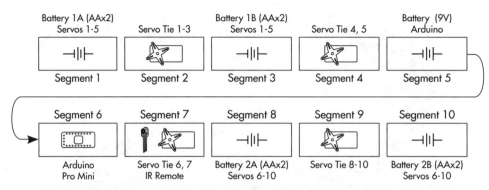

Figure 12-3 Battery power and electronics for the Snakebot are distributed on each segment, as shown. Batteries are placed in AA × 2 battery holders. The battery holders, and homebrew circuit boards, are attached to the Snakebot body using Velcro tabs for easier maintenance.

- A pair of 1" standoffs separates the base and top decks. The standoffs are secured using 4-40 flathead screws. The flat head is important to maintain a smooth surface.

To keep track of the segments they're numbered from 1 to 10. Segment 10 is the head. (Though the Snakebot can be made to travel in either direction.)

ARDUINO BOARD

Snakebot uses an Arduino Pro Mini 328. This is a standard Arduino form factor, but significantly smaller than the Arduino Uno and similar boards you may be familiar with. The Arduino Pro Mini measures just 0.7" × 1.3", and comes on a 28-pin, double-wide board that can be directly plugged into a solderless breadboard.

Be sure to get the 5V version of the Arduino Pro Mini. There's also a 3.3V version, but the 5V is better for the Snakebot. You also want the Pro Mini with the ATmega328 microcontroller chip, rather than the slightly less robust ATmega168. The '328 model supports larger program space.

When compiling and uploading the sketch from the Arduino IDE, be sure to select the correct Mini Pro model. Under Tools | Board choose Arduino Pro/Pro Mini, 5V, with ATmega328.

The Arduino Pro Mini doesn't have pin headers on its I/O pins for connecting up things. You can solder wires or header pins directly to the module. I don't advise soldering wires directly … it's much better to use male or female headers, so you can attach or detach the wiring cables as you need.

You can decide how the pins are soldered to the board—the pins can point down (to plug into a solderless breadboard) or point up (to connect to wiring). My advice: use double-long header pins, which are long on both ends. That way the Arduino Pro Mini can both plug into a breadboard *and* connect with wiring on top. Pretty clever, eh? (Okay, I didn't invent the idea, but it's still a clever one.)

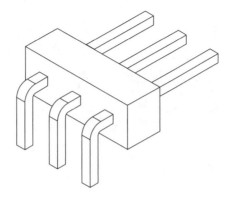

The Arduino Pro Mini has no built-in USB programming adapter, so you need to get one separately. Yes, it's an additional-cost item, but it's a one-time purchase, and can be used with multiple Pro Minis—or most any other Arduino that lacks a USB port for programming. As with the I/O lines, you need to solder on a set of header pins to accommodate the programming adapter. Use a right-angle header.

You have your choice of several forms of programming adapters for the Arduino Pro Mini. One form has the USB electronics embedded in the cable; no additional USB cable is required to connect to your computer. My preference is the breakout board adapter, which you can leave plugged into the header on the Pro Mini. It needs a Mini-B USB cable to attach to your PC. Figure 12-4 shows a USB-to-serial adapter board.

AUTONOMOUS VS. REMOTELY CONTROLLED

Operation of the Snakebot is mostly autonomous—after programming, just connect the batteries, set it on the ground or on a large table, and watch it move. Using an alternate sketch you can operate the snake using a universal TV remote. Control options allow you to activate and deactivate the servos, set speed, and alter the amount of lateral undulation. Both modes are detailed later in the chapter.

Constructing the Snakebot

As noted earlier, Snakebot is composed of 10 segments, all the same. Each segment is composed of three plates: base, servo, and top. The base and servo plates are constructed from 1/4" aircraft-grade plywood or 6mm expanded PVC plastic (I used PVC). The top plates are made from 1/8" plywood or 3mm expanded PVC. In addition, five of the segments also use

Figure 12-4 The Arduino Pro Mini board does not have its own USB connection. The FTDI Basic USB-to-serial board connects between your PC and the Arduino Pro Mini. I prefer this type of breakout board over the all-in-one programming cables, but you can choose whichever you like best.

a duplicate top plate for mounting the Arduino and servo headers—these plates are sandwiched to the top plates on the Snakebot using Velcro tabs.

Refer to Figure 12-5 for the cutting and drilling layouts for all the plates. Dimensions are modestly critical, especially the hole spacing for mounting the servo. Make and assemble one segment to test the fit. Make any adjustments as necessary before cutting and drilling the remaining segments.

You'll cut an extra (11th) base plate that will serve as the head. This segment is a little longer and can accommodate sensors, light effects, small speakers, and other assorted enhancements you may wish to add. These extras are optional, and completely up to you.

Small plastic or wood wheels of 1/2" to 3/4" diameter can be mounted most anywhere along the length of each base plate—check out Figure 12-8 for ideas. I chose to mount the wheels on my prototype Snakebot approximately 3/4" from the end of the baseplate that connects to the spacer. Seemed to work okay.

Now's a good time to mark the holes for the wheels using a spring-loaded center punch. To make attaching the wheels easier, drill a 1/16" or smaller pilot hole into the side of the plastic. Don't attach the wheels just yet.

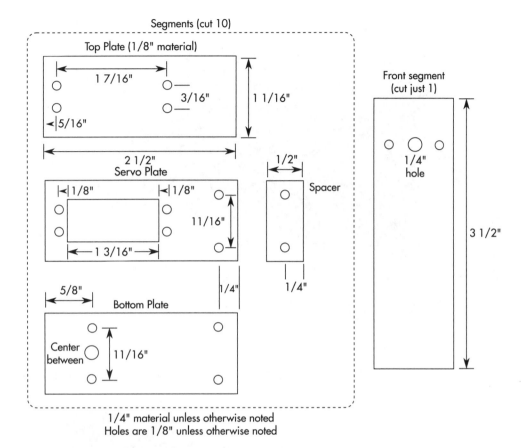

1/4" material unless otherwise noted
Holes are 1/8" unless otherwise noted

Make an additional five top plates without holes for
mounting the tie-point circuit boards and solderless
breadboard to the top of the Snakebot. Use Velcro
tabs to attach these plates to the Snakebot.

Figure 12-5 Drilling and cutting layout for the Snakebot segment pieces. You'll also want to cut
five additional top plates (also using 1/8" material) without the holes. Use these with Velcro tabs
and the circuit board segments; they make it easier to work with the wiring of your Snakebot. (The
fifth additional plate is for the solderless breadboard and Arduino Pro Mini.)

ASSEMBLING THE SEGMENTS

Follow these steps to assemble each segment:

1. Mount a servo to the servo plate using two 1" threaded standoffs and 4-40 screws, as
   shown in Figure 12-6. The underside of the servo plate, with servo attached, will look

Figure 12-6 Construction of a segment, starting with a servo plate, servo, and 1" 4-40 threaded standoffs

like Figure 12-7. The standoffs are a tight fit against the servos. Hint: To make things jibe you'll need to turn the servo wire sideways.

2. Drill mounting holes for screws in the round servo disc that comes with the servo to match the holes in the base plate. Attach the disc to the base plate using 4-40 × 7/16"

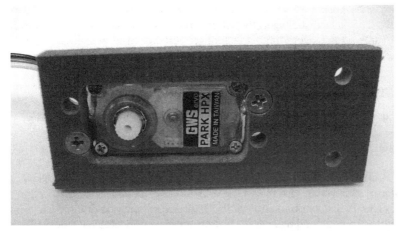

Figure 12-7 Underside of the servo plate, showing the flathead screws used to secure the servo. Notice that the servo is mounted on the back side of the plate.

flathead screws and 4-40 nuts. (If using 1/4" thick material you can probably get by with 1/2" flathead screws, but 6mm PVC plastic will require the slightly shorter 7/16". If you just can't find that length, you can add #4 washers between the plate and servo disc to act as spacers.)

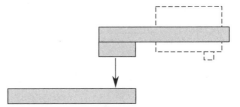

3. Mount a base plate to the servo plate using the plate spacer and two 4-40 × 3/4" flathead screws and 4-40 nuts. See Figure 12-8 for details.

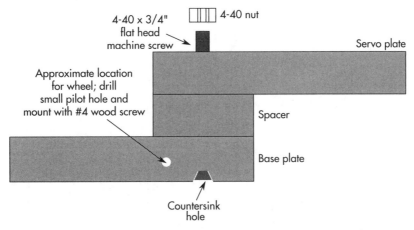

Figure 12-8 Construction detail showing the bottom plate, spacer, and servo plate. Note also the approximate (and suggested) location for holes for the wheels.

Attach the wheels using #4 wood screws. My prototype used 3/4" flathead wood screws, and I placed a 1/8" spacer between the wheel and plastic to keep the wheels from rubbing against their rims. Don't overtighten the screws; you want the wheels to freely turn when rotating. For the wheels themselves you can rob cheap toys from the dollar store.

SETTING THE NEUTRAL POSITION OF THE SERVOS

With the segments complete you can now assemble them to make the snake. However, before doing so you'll want to set the *neutral,* or center, position of each servo. This is an important step to ensure that the body of the snake is straight when all the servos are set to center.

Use the Arduino Pro Mini and the *Snake_ServoCalibrate* sketch that follows to set the neutral center of each servo. Use the wiring diagram in Figure 12-9 for connecting a servo to the Arduino. The 5V regulator aboard the Pro Mini doesn't provide enough current to directly power the servo, so you need a separate battery supply, connected as shown.

 Be extra careful not to reverse the electrical connections to the servo. Doing so many damage the servo, and could also harm the Arduino. The center wire of the servo is +V power. Be sure this connects to the + side of the AA × 4 battery holder.

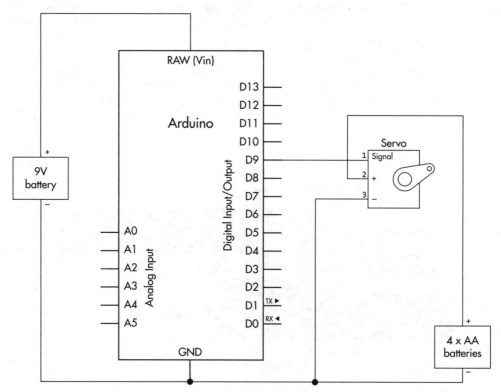

Figure 12-9   Wiring diagram for testing and calibrating each servo, as you assemble the Snakebot segments. Be sure the servos are centered as you put the Snakebot together.

Snake_ServoCalibrate

```
#include <Servo.h>
Servo myServo;

void setup() {
  myServo.attach(9);      // Servo to pin D9 (see text)
  delay(200);

  myServo.write(180);
  delay(1000);
  myServo.write(0);
  delay(1000);
  myServo.write(90);      // Center (neutral) position
}

void loop() {
  // Empty loop
}
```

FYI   Recall from Chapter 4, "Programming the Teachbot: Making It Move," that the Servo library method *attach*—and its opposite *detach*—bind the servo object to an Arduino pin. When attached, pulses are sent to the servo to operate it. When detached, the pulses are removed. In this case, the terms attach and detach don't mean an electrical connection, but a virtual one that controls whether or not the servo is powered.

1. Start with segment 1 (the tail). Connect the servo to the circuit in Figure 12-9 earlier, apply power, and upload the *Snake_ServoCalibrate* sketch to your Arduino. The sketch is written so that the servo first moves all the way in one direction, all the way in the other direction, and finally to the neutral, center position.

2. Keep power applied. Turn the segment on its side, and attach segment 2. Tighten the screw, but be careful not to overtighten. You don't want to strip out anything.

3. Connect the segment 2 servo to the Arduino. Press the reset button on the Arduino to restart the sketch. The servo should move as described earlier. Repeat the assembly steps to attach the segment in the line. Do the same until all the segments are assembled.

4. Attach the 11th (longer) base plate to segment 10, as shown in Figure 12-10.

Keep the snake on its side until all segments have been assembled. The finished Snakebot is a full three feet long, so you'll need to construct it on a wide table, or on the floor.

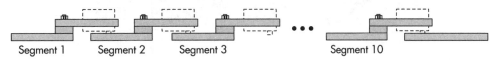

Segment 1   Segment 2   Segment 3   Segment 10

Figure 12-10   How the base plates are mounted back-to-front. Use the longer base plate for the front of the Snakebot—that's segment 10.

PLATES FOR BATTERIES AND SERVO HEADERS

The top plates of segments 1, 3, 8, and 10 each hold a set of two AA batteries, contained in a AA × 2 battery holder. The batteries in segments 1 and 3 are wired in series (see Figure 12-11) to provide 4.8V or 6V for the servos in the rear half of the snake; the batteries in segments 8 and 10 serve power for the servos in the front half of the bot. (For the remainder of this chapter, I'll assume you're using rechargeable batteries; the four cells provide 4.8V.)

The top plates of segments 2, 4, 7, and 9 hold homemade servo headers. These headers act as distribution or *tie-points* for the servo wiring. They reduce the number of long cables running the length of the Snakebot. An example tie-point header board is shown in Figure 12-12. Good

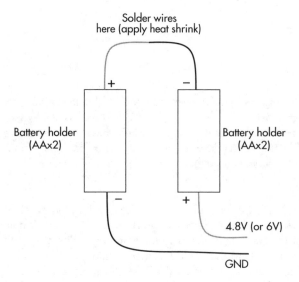

Figure 12-11 The two AA × 2 battery holders on each end of the Snakebot are wired in series, to provide a total voltage of 4.8 or 6V. I recommend using rechargeable NiMH batteries, which deliver 1.2 volts each, or 4.8V for a set of four.

Figure 12-12 Homemade tie-point circuit board reduces the number of wires up and down the length of the Snakebot. The servos from each segment connect to this tie-point; then a single three-wire extension connects to the Arduino.

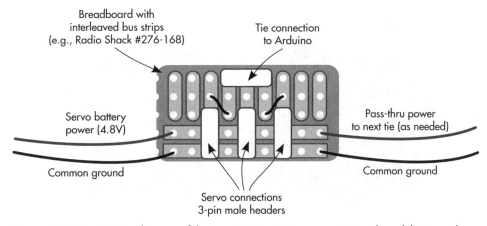

Breadboard with
interleaved bus strips
(e.g., Radio Shack #276-168)

Tie connection
to Arduino

Servo battery
power (4.8V)

Pass-thru power
to next tie (as needed)

Common ground

Common ground

Servo connections
3-pin male headers

Figure 12-13 Wiring diagram of the tie-point circuit. Use a prototyping board that provides for the bus rail pattern as shown. Notice the two jumpers that connect the left and right male header pins to the Arduino tie connection header.

soldering is critical here. If the solder joints are weak they may break as the wires flex during the snake's movement. This will result in intermittent problems with the servos.

Figure 12-13 shows how to construct the headers using bits of cut-up solderboard (RadioShack #276-168; this kind has interleaved long strips with groups of three pins). Before soldering, cut each bit of solderboard to size, and drill mounting holes on either side. The locations of the holes aren't critical, but be sure to leave enough space to attach all the wires. Solder the three-pin breakaway headers as shown. Each set of servo headers provides connection points for up to three servos.

- Power and ground are provided using the long bus rails on the PCB. The top rail carries power from the servo batteries. Note: Don't mix up servo power with the Arduino's 9V power or the regulated 5V provided by the Mini Pro board. The servos derive their juice from the sets of AA batteries on segments 1 and 3, and 8 and 10.
- The bottom rail is the ground connection. All tie-points are connected through this common ground. This is very important. The same common ground also connects the Arduino's GND pin. Your snake won't work correctly if any segment is left unconnected to this common ground.
- The control inputs for the three servos are supplied by a three-pin connector. Use a three-wire servo extension cable to bridge between each tie-point board to the Arduino, as noted in the next section.

Make three tie-points of this size and type. Make a fourth a little wider so you can solder on an infrared remote receiver module, like that in Figure 12-14. Electrical connection is indicated in Figure 12-15. Note that the output signal from the module connects to Arduino pin D11.

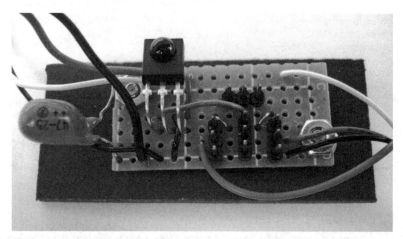

Figure 12-14 The tie-point on segment 7 also includes an infrared receiver module. This circuit board should be a little larger to make room for the module and 47 μF tantalum filtering capacitor.

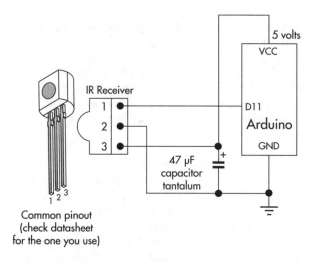

Figure 12-15 Electrical connection for the infrared receiver and Arduino. The output of the receiver is connected to the center tie-point pin on the segment 7 circuit board. This in turn connects to pin D11 on the Arduino.

ADDING SERVO EXTENSIONS

The GWS Park L servos I used for the prototype Snakebot come with fairly short wiring. The wires are just long enough for most, but not all, of the servos to reach their respective header. Short (4 to 6") extensions are needed for the servos on segments 3, 5, and 10.

You'll also need three-wire servo extensions that connect between the servo headers and the pins on the Arduino. You may wish to make these yourself, cut to length, or use premade 12" or 14" extensions. Any extra wire you can neatly bundle up using a small cable tie.

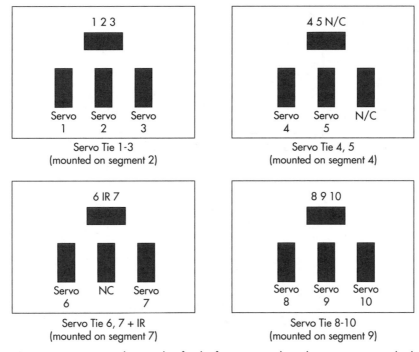

Figure 12-16 Distribution plan for the four tie-point boards. Servos are attached to the three-pin headers as shown. Connect the three-wire servo extension cables to the Arduino (see also Figure 12-19 for how each "bundle" is attached to the Arduino board).

See Figure 12-16 for how the segment servos connect to the four tie-points. Attach the servo extension to the header along the top of each tie-point, and to its corresponding pins on the Arduino (see "Wiring the Arduino Pro Mini" for details).

SEGMENT WIRING

In addition to the extension wiring between servo headers and Arduino, the subsystems on each segment are connected together using ordinary 24- or 26-gauge wiring. Refer to Figure 12-17 for a wiring diagram. Important things to observe:

● A common ground point connects batteries, servo headers, and Arduino. As noted earlier, the common ground is important for proper operation; without this, the servo motors may not operate properly, or at all. (If a servo appears to jitter uncontrollably, inspect its ground connection.)

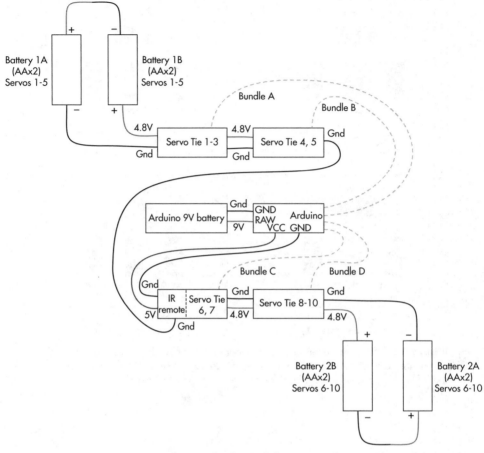

Figure 12-17 Segment wiring diagram for the Snakebot. Notice the ground wire that connects between the battery sources, all segment electronics, and the Arduino. Make absolutely sure this ground connection is not interrupted, or your snake won't move properly—if it moves at all.

- The servos are wired separately from the 9-volt supply for the Arduino, and separately from the regulated 5-volt output from the Arduino. Servo power comes from the two sets of AA × 2 battery holders.
- The servo header on segment 7 includes a self-contained infrared receiver for the optional remote control function.

Be sure to use good soldering techniques on all headers and wiring. During movement of the snake the wires can tug and pull, and any sub-par soldering joints may come apart. All wires should have ample slack so that they do not bind when the segments move. Prevent snags by binding the wiring at several points using cable ties.

If your Snakebot suddenly stops working—the servos just jitter or the thing tends to curl up in a tight ball—suspect a bad connection somewhere. (This behavior can also be caused by drained batteries, so always keep the batteries fully recharged.)

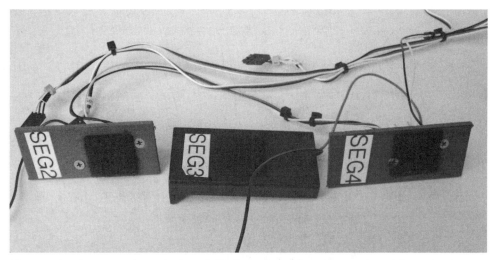

Figure 12-18 Mount the circuit boards for the segment electronics on plates (identical to the top plates, but drill holes to match those in the boards). Then use Velcro tabs to secure these plates—as well as the battery holders—to the Snakebot.

Cut five additional plates out of 1/8" material to the same size as the top plates. Attach the tie plate boards to these plates using 4-40 flathead screws and 4-40 nuts. Mount these boards and the battery holders to the top plates of the snake using small squares of Velcro, as shown in Figure 12-18. This method allows you to keep the snake mechanics and electronics separate to help you work with each separately.

Wiring the Arduino Pro Mini

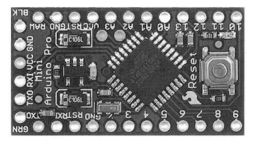

The Arduino is the heart and brain of the Snakebot. As noted earlier in the chapter, I've selected a 5V Arduino Pro Mini for its small size. The Pro Mini is mounted on a 170 tie-point mini solderless breadboard. Electrical connection between the Pro Mini and the board is through two rows of double-long male header pins. These pins are long in both directions, allowing the Pro Mini to plug into the board, yet also provide contact points on top for the servo extension cables.

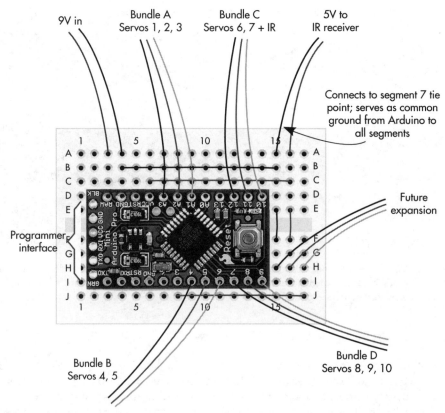

Figure 12-19 Electrical wiring diagram for connecting things up to the Arduino Mini Pro. The "bundles" are from the tie-point circuit boards. See also Figure 12-17.

Consult Figure 12-19 for the layout for the Arduino Pro Mini on its solderless breadboard. All wire connections to the Pro Mini are through female headers, which is how servo extension cables are terminated. Connections to the breadboard are through male headers. Avoid trying to use solid hookup wire for any Snakebot connection. Use only stranded wiring to allow for flexing as the snake slithers across the floor. Attach the Arduino and breadboard to segment 6 using Velcro.

Power to the Arduino comes from a 9V battery mounted (again, use Velcro) to the top of segment 5. Wire the battery to the Pro Mini using a 6" battery cable, terminated in a two-pin male header connector. Attach as shown in Figure 12-19, and keep the battery cable firmly planted to the breadboard with a cable tie.

Use the battery clip as a form of switch. Keep one tab of the clip connected to the battery. When you want to apply power to the Arduino Pro Mini, swivel the clip so the second tab makes contact with the battery.

Of course, this switches power to the Arduino only. The motors have their own battery supplies; to interrupt current to the motors, simply remove one of the batteries from the holders on each end of the snake.

Programming the Snakebot

Programming the Snakebot is essentially a matter of moving its 10 servos in a specific sequence to mimic the sinusoidal undulation of a real-life snake. The sequence has to be just right to achieve forward movement. If the servos aren't synchronized, the snake will just thrash about.

There are several ways to define the servo sequence. One way—and ultimately the best way—is to use a math function to define the sine wave for the undulation. The sine wave looks like the one in Figure 12-20.

- *The period of the wave,* which is the distance between one crest (or valley) to another. For a wave of a finite length, this effectively specifies the number of crests along its length. For undulation locomotion, there should be at least one full period along the Snakebot's body.
- *The amplitude of the wave,* which is the amount of variation between crest and valley. Given a low amplitude, there will be little variation in the undulating motion; a high amplitude is just the opposite (see Figure 12-21). With a high amplitude the snake will travel farther for each undulation. Because the Snakebot is like a string of a specific length, the higher the amplitude of the wave, the shorter the maximum period.
- *The frequency of the wave,* which is how fast the wave changes from crest to valley. The higher the frequency, the faster the wave will change. For the Snakebot this means the faster it will move.

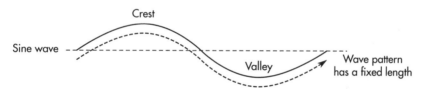

Figure 12-20   The Snakebot moves by undulating its body in a sine wave. Parameters of this sine wave influence the speed and other factors of the snake motion.

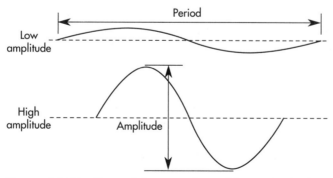

Figure 12-21   The Snakebot moves faster if the amplitude of the sine wave is increased. Speed is likewise increased by moving servos faster.

To make the Snakebot sketch easier to visualize and follow, and to allow empirical study, I've elected to hard-code the sine wave function. This includes preformatting the sine motion in a 12-element byte array—using an array means you can easily alter the motion parameters to experiment with different locomotion techniques without having to apply any math functions.

Waveform amplitude and frequency are set in the Snakebot programming using variables that are initialized at the top of the sketch (more about this in a bit). Frequency is determined by the *transit speed* of the servos—the time it takes for the servo to rotate from one position to another.

Ordinarily the transit speed is determined by the motor and gearing inside the servo, and it's generally desirable to make the transit as fast as possible. But for the Snakebot the transit speed must be slowed down, necessary not only to provide a smoother motion, but also to make it possible to alter the frequency of the sine wave.

Slowing the servo transit time is the most involved aspect of Snakebot's programming. Understanding the available methods, and how it's handled in the Snakebot, is important to properly adjusting the sketch to experiment with the best locomotion. So I'll describe that first, and then discuss the full sketch.

SLOWING SERVO TRANSMIT TIME

So again, in normal circumstances you want the fastest servo motors possible—the quicker the servo can move from one end of its arc to the other, the better. But that's not what we want for a robotic snake. With fast servos the undulating motion of the snake is too abrupt, which could cause poor locomotion. By slowing the speed of the servos your robotic snake moves more gracefully, and with less thrashing—that means with more efficient motion.

There are a couple of ways of slowing down a servo:

Supply the servos with a lower voltage. Like most motors, servos move more slowly when powered with fewer volts. But there are several gotchas with this approach: The biggest problem is that the lower the voltage, the lower the torque—the motor is weaker and not able to move or push the same weight.

Use a separate servo module with a speed control function. A number of third-party servo motor control products incorporate a speed setting feature. You can use it to run the servos at a slower-than-usual speed.

Control servo speed using the Arduino. By moving the servo in small increments instead of one complete motion, you can insert short timing delays that effectively slow down the sweep. It's this last method that we'll use in the Snakebot, and the one I'll discuss in this chapter.

The Arduino Web site provides an example of incrementing a servo motor with delays. Check out the *Sweep* sketch on the example page for the Servo object. *Sweep* moves a servo from 0 to 180, in one-degree increments, and with 15 milliseconds (ms) delay between each degree.

This approach is perfectly acceptable, but the use of the *delay* statement makes it inefficient, and not easily scalable to handle lots of servos. Every time there's a *delay* statement the sketch literally freezes there for the allotted time, and only then continues on.

Rather than delaying with *delay,* you can delay the incremental motion by using a hardware timer. This timer, which is like an alarm clock, runs in the background of the Arduino. Each time the timer goes off, it triggers the next increment. While this may sound complicated, the technique is not difficult to implement, and the code is clean and compact. It's easy to remove or add servos. More about this technique in the next section.

Two separate timers—using two different techniques—control servo speed. These timers are:

- A *software-based timer,* implemented using the *Metro* timer library (available on the *Arduino.cc* Libraries page, and provided for your convenience on the ARB Support Site detailed in Appendix A), defines how often the sine wave makes a complete cycle. One cycle is not the same as one period: a cycle involves the motion of all 10 segments, which is about 1.5 periods. This timer is updated roughly once every half-second.
- A *hardware-based timer,* implemented using the *MsTimer2* library (also available on *Arduino.cc* and the ARB Support Site), controls the transit speed of the servos. Each "tick" of this timer, which occurs every 40 milliseconds (ms), updates the position of the servos.

MsTimer2 uses one of the hardware timers (specifically Timer2) built into the Arduino. Being hardware based, there can only be one *MsTimer2* timer set at a time. Whenever the timer goes off it generates an *interrupt,* which in turns calls a function, *moveInterval.* Inside this function:

A *for* loop cycles through all 10 servos. Each servo is referred to as an element in an array called *servo.* In this way the 10 servos can be referenced using the control variable in the *for* loop.

For each servo in turn, the current position of the servo is compared against its *endpoint,* the final position where the servo is headed. Endpoints are stored in an array called *segments.* This array, defined near the top of the sketch, has 16 elements. That's enough for all 10 servos to move through 1.5 periods of the sine wave.

If the endpoint is greater than the current servo position, the servo is incremented forward. Conversely, if the endpoint is less than the current servo position, the servo is decremented backward.

Here's the *moveInterval* function for study:

```
void moveInterval() {
  for (int i = 0; i < 10; i++) {        // Go through all 10 servos
    if (segments[seq+i] > servo[i])     // Endpoint > cur position
      servo[i]+=stepAmount;             //   servo goes forward
    else if (segments[seq+i] < servo[i]) // Endpopint < cur position
      servo[i]-=stepAmount;             //   servo goes backward
    else servo[i] = segments[seq+i];
    myServos[i].write(servo[i]+90);     // Move servo to new position
  }
}
```

The Snakebot sketch defines important variables that alter the behavior of the sine wave undulation of the servo motors:

```
const int stepInterval = 40;
const byte stepAmount = 2;
const int seqInterval = 550;
const char maxV = 45;
const char minV = -45;
```

- *stepInterval* (in milliseconds) is the time delay between each incremental movement of the servos. The shorter this delay, the faster the servos move to their next step.
- *stepAmount* (in degrees) is how far the servos move at each interval. The lower the number, the less the servos move during each interval. This value should be kept low, usually just 1 or 2.
- *seqInterval* (in milliseconds) is the time delay between each cycle. A cycle is a sequence of six *waypoints* that the servos travel through that define the sine wave. A continuous sine wave is produced by indefinitely repeating these six waypoints, offset by one for each segment in the snake.
- *maxV* and *minV* set the amount of deflection of the servos on both sides of the center point. The higher these numbers, the greater the amplitude of the sine wave. For basic locomotion both *maxV* and *minV* should be the same.

The values of these five variables play upon one another; change one and you need to adjust the others. For example, when increasing the *maxV* and *minV* deflections you need to allocate either more time for the *seqInterval,* or increase *stepAmount* to move the servos a greater distance each time they are incremented. I won't go through all the permutations here; feel free to experiment to find your own optimal settings.

LOADING AND RUNNING THE SNAKEBOT SKETCH

The *Snake_ServoSegmentMove* sketch demonstrates how the 10 servos of the Snakebot can be moved in sequence to produce a basic sine wave locomotion pattern. In preparation for uploading and running the sketch, insert three of the four *freshly charged* AA batteries for each set of servos. Leaving one out prevents the servos from moving when you don't intend them to.

Attach the USB upload cable and programmer to the Arduino Pro Mini. Temporarily change the two lines near the top of the sketch to the following:

```
const char maxV = 10;
const char minV = -10;
```

As noted earlier, the *maxV* and *minV* values reduce the amount of deflection of the servos in case there's an issue with construction or wiring.

Compile and upload the sketch. After loading, connect the 9V battery, *depress and hold* the reset button on the Arduino, add the remaining batteries to the holders, and place the Snakebot on the floor. Release the reset button, and the Snakebot should then move.

Remember to use freshly charged batteries with the Snakebot. Each set of four batteries must operate five servos; that's a lot of current demand, and the batteries can drain quickly. As the batteries drain, their voltage drops, and the servos may not operate properly. If the servos of your snake are misbehaving—they jitter or just don't go in the direction they're supposed to—recharge the batteries and try again.

Upon a successful first test (the snake undulates, and may even move forward a bit!) revise the *maxV* and *minV* values back to their defaults:

```
const char maxV = 45;
const char minV = -45;
```

Recompile and reupload.

The *Snake_ServoSegmentMove* sketch has a calibrate mode, should you ever need to realign the servos to their center neutral position. Change the line

```
#define calibrateServos false
```

to

```
#define calibrateServos true
```

You can also use the *calibrateServos* setting when you want to test the function of the servos but don't want the Snakebot to move.

Snake_ServoSegmentMove

```
#include <Metro.h>                      // Segment motion timer
#include <MsTimer2.h>                    // Servo speed timer
#include <Servo.h>

#define calibrateServos false

const int stepInterval = 40;            // MsTimer2 interval (ms)
const byte stepAmount = 2;              // Degree spacing
const int seqInterval = 550;            // Metro timer interval (ms)
const char maxV = 45;                    // CCW deflection
const char minV = -45;                   // CW deflection

const char segments[] = {                // Segment array
  minV, minV, 0, maxV, maxV, 0,         //   (enough for 10 servos,
  minV, minV, 0, maxV, maxV, 0,         //   6-step sequence)
  minV, minV, 0, maxV, maxV, 0 };

Metro serialMetro = Metro(seqInterval); // Set up segment motion timer
Servo myServos[10];                      // Set up 10 servos
```

```
volatile byte seq = 0;                        // Current sequence (out of 6)
char servo[10];                               // Current position for 10 servos

// Main setup function
void setup() {
  myServos[0].attach(A1);      // segment 1 // Attach all 10 servos
  myServos[1].attach(A2);      // segment 2
  myServos[2].attach(A3);      // segment 3
  myServos[3].attach(6);       // segment 4
  myServos[4].attach(5);       // segment 5
  myServos[5].attach(10);      // segment 6
  myServos[6].attach(12);      // segment 7
  myServos[7].attach(9);       // segment 8
  myServos[8].attach(8);       // segment 9
  myServos[9].attach(7);       // segment 10 (head)

  // Either calibrate servos, or run motion
  if (!calibrateServos) {
    // Bind MsTimer2 to 'moveInterval' function at stepInterval
    MsTimer2::set(stepInterval, moveInterval);
    MsTimer2::start();
  } else {
    for (int i = 0; i < 10; i++) {
      myServos[i].write(90);
    }
  }
}

// Main repeating loop function
void loop() {
  if (!calibrateServos) {              // Not in calibrate mode
    if (serialMetro.check() == 1)      // If Metro timer has elapsed
      moveSegment();
  }
}

// Increment through 6 segment states
void moveSegment() {
  if (seq < 5) {   // If current segment < 5
    seq++;         //    go to next
  } else {         // Current segment > 5
    seq = 0;       //    reset to 0
  }
}

// Set slow speed of servo; iterate through all 10
void moveInterval() {
  for (int i = 0; i < 10; i++) {              // Iterate for 10 servos
    if (segments[seq+i] > servo[i])           // Increase CCW to setpoint
      servo[i]+=stepAmount;
    else if (segments[seq+i] < servo[i])      // Decrease CW to setpoint
      servo[i]-=stepAmount;
    else servo[i] = segments[seq+i];

    myServos[i].write(servo[i]+90);           // Normalize for 0-180 degrees
  }
}
```

On the Web: Using the Arduino as a Servo Controller

You can always use a separate Arduino as a custom servo control module. This frees your main Arduino controller to do the other robotic tasks you need, as there's no longer a requirement to do everything in one chip. The two Arduinos are linked via a serial communication connection—the main *host* Arduino sends simple commands to the second Arduino acting as servo *co-processor*. These commands specify the servo to control its position and the transit speed. You can control up to 12 servos when using an Arduino Uno or similar development board that is based on the Atmel ATmega328P microcontroller.

As an exercise, and knowing I'd have a use for it someday, I put together an Arduino sketch—I've named it *ArduinoSerialServoControl*—that listens to the hardware serial port for servo position commands that you send from another Arduino (or, for that matter, other microcontrollers). As the sketch is not directly relevant to this chapter, I've put it on the ARB Support Site (see Appendix A) as a free bonus project.

The *ArduinoSerialServoControl* sketch handles up to 10 servos connected to Arduino pins D3 through D12. The remaining pins are available for other tasks. See the ARB Support Site for programming examples, but in general terms:

From the host controller send the following five elements; this sets one of the (up to) 10 servos:

```
w.s.[servonum].[pos].[delay]
```

- The *w* and s are letter characters that tell the servo co-processor to "write" to a servo.
- *[servonum]* is the number of the servo, from 0 to 9, that you wish to control.
- *[pos]* is the position, from 0 to 180, of the servo. Recall from Chapter 4 that these values represent a servo angle position, in degrees, with 90 being the center.
- *[delay]* is the amount of delay, in milliseconds, between one-degree increments. Setting the value to 0 means no delay.

The five elements are each separated by a period. Here's an example:

```
w.s.2.120.2
```

means move servo number 2 (it's connected to pin D5) to position 120, with a 2ms delay between one-degree increments.

Operating the Snakebot by Remote Control

The infrared receiver mounted on segment 7 allows you to use a universal remote control to operate the Snakebot. The remote function uses the *IRremote* library by Ken Shirriff—see the ARB Support Site (Appendix A) for download locations.

With the *IRremote* library you can communicate with your Snakebot using an ordinary TV/VCR universal infrared remote. Buttons you press on the remote are received by the Arduino and interpreted by the sketch in whatever way you like.

Figure 12-22   Use a TV remote control to experiment with the behaviors of the Snakebot's sine wave. The number buttons change the three main parameters of the sine wave; pressing number 0 activates and deactivates the Snakebot's motors.

For the *Snake_RemoteControl* sketch available for download from the ARB Support Site, the following buttons are mapped to the control variables used to define speed and servo deflection (see Figure 12-22):

- The 0 button starts the Snakebot (the servos are enabled).
- Pressing 0 again stops the Snakebot (the servos are disabled).
- The 1 and 7 buttons increase/decrease the *stepInterval*, which affects the speed of undulation.
- The 2 and 8 buttons increase/decrease the *seqInterval*, the delay between cycles.
- The 3 and 9 buttons increase/decrease the *maxV* and *minV* (same value for both), which control the servo deflection.

Snake_RemoteControl is "tuned" to work with Sony TV codes; most universal remote controls support this code set. Check the instructions sheet that comes with the remote for how to program it for a Sony TV. On most universal remotes the code is a three-digit number, like 098 or 114—you may have to try several codes to get one that works. Use the *Snakebot_RemoteTest* sketch to verify proper operation of your remote.

Most universal remotes are multi-function, meaning they support TVs, VCRs, DVD players, and other equipment. Be sure to select the TV mode in order to match the Sony TV code you've entered. Otherwise, pressing buttons on the remote will have no effect. Sony TV codes are distinct from Sony VCR and other types of codes.

To use, upload the sketch, then open the Serial Monitor window. The buttons you press on the remote should be echoed in the Serial Monitor window. Number buttons 1 through 9 are mapped.

Snakebot_RemoteTest

```
#include <IRremote.h>          // Include IRremote library
                               // Put in sketchbook /libraries folder

#define  showCode  false       // Test mode
                               // false=match against number buttons
                               // true=show raw code

const int RECV_PIN = 11;       // Receiver input pin on 11
IRrecv irrecv(RECV_PIN);       // Define IR receiver object
decode_results results;

void setup() {
  irrecv.enableIRIn();         // Start the receiver
  Serial.begin(9600);
}

void loop() {
  if (irrecv.decode(&results)) {  // If valid value was receiver
    if(showCode) {                // If showCode=true
      Serial.print("0x");
      Serial.println(results.value, HEX);  // Show raw hex value of code
    } else {                      // else showCode=false
      switch (results.value) {    // Match button against Sony TV codes
        case 0x10:
          Serial.println("1");
          break;
        case 0x810:
          Serial.println("2");
          break;
        case 0x410:
          Serial.println("3");
          break;
        case 0xC10:
          Serial.println("4");
          break;
        case 0x210:
          Serial.println("5");
          break;
        case 0xA10:
          Serial.println("6");
          break;
        case 0x610:
          Serial.println("7");
          break;
        case 0xE10:
          Serial.println("8");
          break;
        case 0x110:
          Serial.println("9");
          break;
      }
    }

    irrecv.resume();                // Receive the next value
    delay(10);                      // 10ms delay
  }
}
```

The *Snakebot_RemoteTest* sketch has two modes, set by the *showCode* constant near the top:

- When set to *false,* the sketch displays the number of the button you press: press button 1, and a "1" appears in the Serial Monitor window.
- When set to *true,* the sketch displays the raw code value from the remote. The value is shown in hexadecimal (base 16) format. The number begins with a *0x*—this is the prefix the Arduino uses to denote hex numbers. For example, the value might be *0x910*; the *0x* means the *910* number that follows is hexadecimal.

 Universal remote controls are a common find at discount stores. You don't need a fancy one, but the higher-quality remotes put out more infrared light, and therefore work over longer distances. Try another remote if the range seems to be limited—you want consistent reception when the remote is at least six feet from the receiver module on the Snakebot. To maintain a good range, be sure the batteries in the remote are fresh.

Once your remote checks out, upload the *Snake_RemoteControl* sketch to the Arduino. When the sketch starts the servos are initially detached (servos deactivated), so they don't move. Press the 0 (zero) button on the remote to attach (activate) the servos. Press it again to turn the servos off.

Experiment with different timing delays and servo deflections to see what effect they have on the locomotion of the Snakebot. The remote control is a great way to play with the setup values to see how variations contribute to—or hinder—motion.

Snake_RemoteControl

 The *Snake_RemoteControl* sketch is too lengthy for printing here. Download it from the ARB Support Site; see Appendix A for details.

Chapter 13

Robby Armstrong

Just because it doesn't move, it doesn't mean it's not a robot. Some of the world's most sophisticated bots are nailed to the floor. They're robotic arms, used for such jobs as building cars, playing a mean game of chess, and even delicate surgery. The robot arm is a science unique to itself, with its own special set of challenges, techniques, and solutions.

In this chapter you'll learn how to build an extensible—you can build onto it—Arduino-based arm. We'll call our *arm-o-bot* Robby Armstrong, a salute to both Robby the Robot from the classic film *Forbidden Planet,* and the 1970s kids' toy Stretch Armstrong, a rubberized wonder without a backbone, but with amazingly muscular biceps.

The completed Robby Armstrong, with separate operating pendant and LCD display panel, is shown in Figure 13-1. Robby has a reach of about 16 inches, including its twin-finger gripper, and can lift a couple of ounces.

 Be sure to visit the ARB Support Site (see Appendix A) for code examples and variations. All sketches from this book are available for ready download, including those that are too long to fit in this book.

Understanding Robotic Arms

Robotic arms come in many forms. Some are modeled after the human arm, with a shoulder, elbow, and wrist. Others are designed more like forklifts. Robotic arms are classified by their "reach," or more properly, the shape and extents of their *work envelope.* This work envelope is defined by a *coordinate system*; the human arm is said to have *revolute* coordinates. The forklift bot, with a swiveling arm that goes up and down, has *cylindrical* coordinates.

Figure 13-1 The completed Robby Armstrong, complete with two-finger gripper, joystick and keypad control, and LCD display

As with human arms, movement in a robotic arm is provided by one or more *joints,* also called *articulations.* Each joint provides a *degree of freedom* (*DOF*); the more degrees of freedom, the more elaborate the reach of the arm. The basic degrees of freedom in a robotic revolute coordinate arm are shown in Figure 13-2.

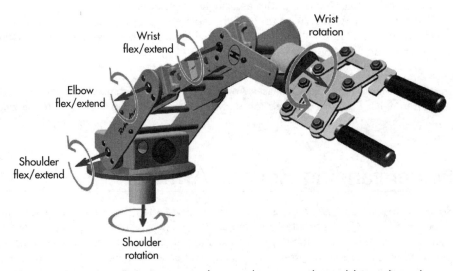

Figure 13-2 Common joints in a revolute coordinate arm. The model is similar to the human arm, but the "shoulder" is a rotating base.

While adding more DOFs provides for a greater range of motion, the cost and complexity of construction increases exponentially. Also, each joint needs a motor or some other mechanism, which increases weight. The more the arm weighs, the less it can lift. At some point, the arm may become so heavy that it can't even raise itself.

A revolute arm with three degrees of freedom provides shoulder and elbow motion. An arm that can't grasp at anything isn't too useful, so a gripper—called an *end-effector* in robotics parlance—completes the arrangement.

Dissecting Robby

 See the ARB Support Site (Appendix A) for a full parts list for building Robby. Included are numerous online sources to help you collect all the pieces.

Robby is a six-DOF revolution coordinate arm. It uses one R/C servo motor for each joint. Construction materials are 1/4" plastic (ABS or high-density polypropylene preferred) and aluminum extruded U-channel.

Additional stamped metal servo connector parts are used to simplify several of the joints. Figure 13-3 shows an example of one way some of these parts may be combined with ordinary R/C servos to provide different kinds of mechanisms. The stamped parts are available from Lynxmotion, Pitsco, and several other online resources. To finish Robby you'll need the

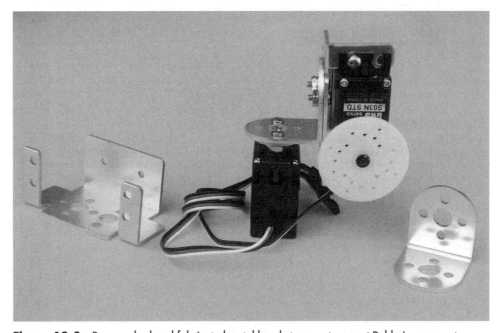

Figure 13-3 Pre-punched and fabricated metal brackets serve to mount Robby's servo motors. Available from several online sources; Lynxmotion calls their brackets Servo Erector Set.

following metal servo brackets. For your reference, part numbers for both Lynxmotion and Pitsco are provided.

Qty	Description	Lynxmotion Part #	Pitsco Part #
2	Multipurpose single servo bracket	ASB-04	W39060
2	L-bracket	ASB-06	W39063
2	Long C-bracket	ASB-05	W39062

Robby's main points of interest are:

- A *bottom plate* provides a mounting for all the parts. The arm is anchored to this plate.
- A *revolving base* provides shoulder rotation. The weight of the arm is supported by a 3" "Lazy Susan" ball-bearing turntable. The base is not securely fastened to the foundation plate; instead, it's made to "float" so that when the arm rotates, any slight misalignment of the pieces does not cause it to bind.
- An *upper arm and forearm* provide shoulder and elbow motion. Both arm pieces are constructed out of short pieces of 1" by 9/16" aluminum U-channel. (Note: The actual dimensions of the U-channel used for the prototype Robby are more like 7/8" × 9/16", but you can use another size as long as it's more or less similar.)
- *Flexing and rotating wrist,* using a pair of servo motors contained in metal servo connector parts.
- *Gripper* attached to the wrist; fingers on the gripper open and close to grasp objects. This chapter does not detail the construction of the gripper, though if you wish to build it, plans are provided as a bonus project on the ARB Support Site. You may find a commercially made gripper kit is easier, and not much more expensive, than building one yourself from scratch. You can choose most any gripper style that will fit the end of the wrist.

All six servos of Robby Armstrong are controlled by a single Arduino. As a demonstration of the functionality of the arm, the servos are controlled using a pair of five-position switches—they act as joysticks—or a 12-key keypad. A two-line LCD provides position feedback.

DEGREES OF FREEDOM

Recall that each joint of a robotic arm is a separate degree of freedom. Throughout the remainder of the chapter the joints of Robby Armstrong are referred to as the following:

Joint	Abbreviation
Base	bse
Shoulder	shl
Elbow	elb
Wrist flex	wfx
Wrist rotation	wrt
Gripper	grp

The three-letter abbreviation is used to display the joint name in the LCD display.

SERVO SELECTION

Robby Armstrong needs the right servos to work properly. Critical aspects of these servos include:

- Standard size, which is more or less 20mm × 40mm × 28mm (these dimensions measure the body of the servo, and do not include the mounting flange). This allows the servos to fit the standard metal and plastic brackets used in the construction.
- Higher torque motors are needed for several of the joints, particularly the shoulder and elbow. In Table 13-1 the servos are graded by torque; all are rated at the 6 volts provided by the servo power. Medium torque: 60–75 oz-in; high: 130–140 oz-in; ultra: 175–185 oz-in.
- Both the base and high-torque servos should likewise use ball-bearing construction. Metal gears are preferred for longer life.

 Servos that can handle 7.2-volt operation are handy, though not absolutely necessary. The higher voltage allows for slightly better torque. Some servos rated for 4.8 to 6V can be operated at 7.2 volts, but may have a shortened life.

SERVO HORN DISCS

Servo horns are the plastic (sometimes metal) attachments that lock onto the shaft of the servo and provide a mechanical link to other parts. Robby uses two styles of servo horns: large round disc and small round disc. Many Futaba and Hitec servos come with these horn discs, but if yours doesn't,

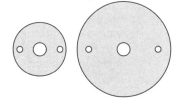

Table 13-1 Servo Selection Guide			
Servo	**Torque**	**Ball Bearing**	**Digital****
Base	Medium	o	
Shoulder*	Ultra-high	o	o
Elbow	High	o	o
Wrist flex	High		
Wrist rotation	Medium		
Gripper	Medium		

* As an option you can use a 1/4-scale servo, such as the Hitec HS-755HB, for the shoulder. This motor is larger and stronger than the average standard-size servo. It may actually be less expensive than an ultra-high torque standard servo, but it does require a change in the bracket used to attach it to the rotating base. If using a 1/4-scale servo, you must replace the metal brackets used for the shoulder joint. Larger versions of the multipurpose bracket and long C-bracket are available. Also consider adding mechanical dampener discs (Lynxmotion part #MD-01), which smooth out the motion of the larger servos.

** Recommended, but not strictly required.

you'll need to get them separately. They're not expensive, and you don't need the metal variety, so save your money.

To mount the discs you need to drill out two 1/8" holes on opposite sides. A drill press makes this job easier. Avoid holding the disc with your bare hands while you drill; grab it using a pair of locking pliers. Don't squeeze down too much on the pliers, or you'll deform the plastic.

SPECIALIZED ARDUINO

Robby uses a special version of the Arduino board called the *Romeo*. It's made by DFRobot, and similar boards are available from other companies. Romeo is particularly well suited for robotics applications, as it has male pin headers for easy connection to servos, switches, and other electronics.

When shopping for compatible Romeo boards, make sure it uses the Atmel ATmega386 chip, and not the smaller capacity ATmega168. Follow the directions that come with the board on how to choose the proper Arduino type when compiling and uploading. Many are programmed as an "Arduino Duemilanove w/ ATmega328."

While the prototype Robby Armstrong used a Romeo board, there are other variations on the theme you can consider. For example, you can use an Arduino form factor board with a servo shield attached. These, too, provide a separate power connection for the servos.

You can also wire up your own headers for the servos and other electronics. Use a solder prototyping board to construct a series of six 3-pin headers for the servos. Some important implications:

- The servos must be powered from a separate source from the Arduino. In Robby, the Arduino (and other electronics) is powered by a single 9-volt battery. The servos are powered by a set of rechargeable C battery cells.
- The ground connection for the servo and servo batteries must be shared with the ground for the Arduino battery.
- It's a good idea to add one or two large filtering capacitors (47 µF to 330 µF) between the servo V+ and ground connections. This helps smooth out any power glitches caused by the servos. Make sure the capacitors are inserted with the proper polarity. You may use either tantalum or aluminum electrolytic capacitors.

Constructing Robby Armstrong

Robby Armstrong uses metal and plastic parts, some homemade and some store bought. Refer to the ARB Support Site (see Appendix A) for a complete parts list with sources.

Robby Armstrong is designed to use as much homebrew construction as possible, not only to save money, but to help foster a complete DIY experience. This doesn't mean building everything yourself is necessarily the best route (only you can decide that), or even the most economical, when also considering your investment in time.

Before starting construction, you may wish to consider augmenting or replacing one or more homebuilt components of Robby with commercial versions. This is especially true of Robby's rotating base, which uses a floating design to allow for a margin of error in construction. The float also makes the base a little less accurate and stable.

You may opt to use a commercially built base, such as the Lynxmotion BR-KT (see Appendix A for a list of sources for parts). This kit comes with a servo, support bearings, and other mechanical parts to construct a fully functional rotating base.

PRE-CENTER ALL SERVOS

Before construction use your Arduino to set the center position of all servos. Use the following *Servo_Center* sketch to set the center position. Connect a servo to pin D4 and run the sketch. The servo will move to its center (1500 µs) position and stop. Repeat for all the servos.

Servo_Center

```
#include <Servo.h>
Servo calibrateServo;
const int servoPin = 4;              // Connect servo to D4
void setup() {
  calibrateServo.attach(servoPin);
  calibrateServo.writeMicroseconds(1500);
}

void loop() {
  // Empty loop
}
```

CUTTING THE BOTTOM PLATE

The base plate provides a platform for mounting the arm mechanics and electronics. Only the four holes for the rotating base are shown. The hole placement corresponds to the mounting holes in a 3" ball-bearing turntable ("Lazy Susan") mechanism. On the one I used, the holes were placed in a square pattern about 2-9/16" apart, as shown in Figure 13-4.

Use the turntable mechanism itself to mark the four holes in the bottom plate at the approximate location shown. The space to the right and left are for mounting the electronics and batteries. Round or lop off the corners of the bottom plate so that they won't snag on anything.

You'll be drilling more holes into the bottom plate as we go along. Hole location is made by marking the actual parts that are mounted on the plate. So get your hand-held power drill ready!

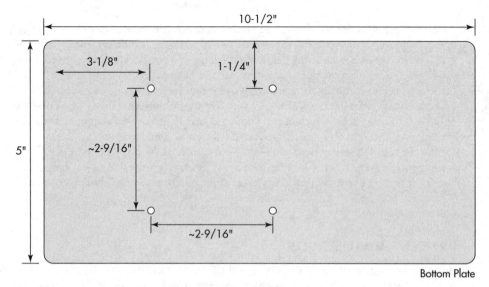

Figure 13-4   Basic drilling and cutting guide for the bottom plate. The holes match those in a 3" ball bearing- ("Lazy Susan") turntable. Other holes you will drill as you go.

BUILDING THE SHOULDER BASE PARTS

Figure 13-5 shows the basic drilling and cutting guide for the plastic shoulder parts. You can use 1/4" aircraft-grade plywood or a plastic. I opted for 1/4" high-density polyethylene,

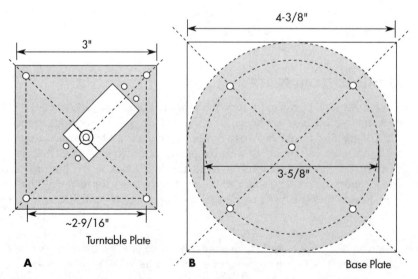

Figure 13-5   Drilling and cutting template for the turntable and base plate. See text for locating the mark for the center hole. The four perimeter holes match those in a 3" ball-bearing turntable.

otherwise known as HDPE. This isn't the kind of stuff you're apt to find locally, but it's available online (see the ARB Support Site). You'll need two 12" × 8" squares.

Note that only some of the holes have a specific drilling location. Measuring and drilling accurately placed holes is difficult without specialized tools and skill. It's often better (not to mention easier) to use the actual parts for marking holes for drilling. When hole placement is shown with a ~ (tilde) character it means approximate location; mark the holes you drill with the actual part that fits it.

Cutting and Drilling the Turntable Plate

The turntable plate has corresponding holes for the turntable mechanism, plus a cutout for the base servo.

1. Use a piece of graph paper to accurately mark the dimensions of the turntable plate. Use low-tack glue to mount the paper to the plastic sheet.
2. Using the turntable mechanism for the exact hole placement, mark and drill the four 3/16" mounting holes in the corners.
3. Use the lines on the graph paper to find the exact center of the plate by drawing two intersecting lines, like that in Figure 13-5a. Mark and drill a 1/8" pilot hole in the center.
4. Cut out the plate.
5. Enlarge the center hole to 1/4", and use this hole as a guide to mark the outlines and mounting holes for a standard servo. Do this by inserting the shaft of the servo into the hole and positioning it at a 45-degree angle, as depicted in the illustration. Use a pencil to mark the outline of the servo, as well as the mounting holes. (You can mark just two holes, on opposite corners.) Remove the servo from the hole.
6. Drill the mounting holes for the servo.
7. Make a cutout for the servo by inserting a coping saw blade through the 1/4" hole you made in step 5 and carefully cutting through the outline.

Cutting and Drilling the Base Plate

The base plate holds the arm and allows it to rotate over the turntable mechanism. This base is fashioned from a 4-3/8" square; you can cut out the base to make a circle like that in Figure 13-5b, or you can merely lop off the corners and make an octagon shape.

1. As with the turntable base, use a piece of graph paper to plot the dimensions for marking and drilling. Draw out the overall square shape of the plate, and find the exact center by marking two intersecting lines. Don't drill it quite yet.
2. Use a drafting compass—the kind with the pencil on one end and needle point on the other—to mark a 3-5/8" diameter circuit from the center.
3. Mark four holes along the intersections of the circle and the previously drawn lines. These holes should correspond to those in the four corners of the turntable mechanism. Check to be sure.
4. Drill the center and corner pilot holes with a 1/8" bit.
5. Go back and enlarge the corner holes (but *not* the center hole) to 3/16". The larger holes provide a bit more leeway for alignment errors.

 Though you'll mark and drill four corner holes, you'll only be mounting the turntable to two of them.

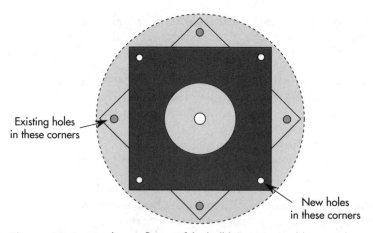

Existing holes
in these corners

New holes
in these corners

Figure 13-6   Use the top flange of the ball-bearing turntable to mark four "sight" holes, which are used to access the mounting screws to the turntable.

Add four "sighting holes" in the base plate by temporarily attaching the turntable mechanism to the plate—use just two screws in opposite corners for this. Finger-tighten only. Then:

1. Turn the plate over so that the turntable is facing up.
2. Turn the turntable so that its top flange is at a 45-degree offset from the flange below. Use the top flange to mark the position of four sighting holes, as shown in Figure 13-6. These holes will allow you to access the screws used to attach the base plate to the turntable plate. After marking, remove the turntable mechanism, and drill the holes with a 1/8" bit.
3. Enlarge all perimeter holes to 3/16". Then enlarge the sighting hole(s) to 1/4".

ASSEMBLING THE SHOULDER BASE

With the shoulder base parts cut and drilled you can now assemble them. You'll need the following fasteners:

Quantity	Fastener
8	4-40 nylon locking nuts
6	4-40 × 7/16" flathead screws
8	4-40 × 1/2" panhead screws
2	4-40 × 3/4" flathead screws
4	4-40 × 3/4" panhead screws
4	1" 4-40 threaded aluminum standoffs
8	4-40 nuts (standard, no locking)
2	1/8" long #4 spacers (or stack of #4 washers)
4	#4 washers
2	#4 external tooth lock washers

Mount Servo Disc to Base Plate

1. Use a 1/8" bit to drill two holes in opposite sides of a large servo disc (if the disc does not come with the servo you'll need to get one separately).
2. Use a 4-40 × 1/2" screw and nut to temporarily attach a large servo disc to the center hole in the base plate.
3. Using the holes you drilled in step 1, mark corresponding holes in the base plate. Drill with a 1/8" bit.
4. Remove the servo disc and drill out the center hole to 1/4".
5. Permanently mount the disc to the plate using two 4-40 × 1/2" screws and nylon locking nuts. Tighten the nut, then back off about 1/8 to 1/4 turn so that the servo disc can float over the center. The base plate with servo disc mounted underneath should look like Figure 13-7.

Mount Servo and Turntable to Turntable Plate

1. Use 4-40 × 7/16" flathead screws and standard nuts to mount the servo to the turntable plate. You can use just two screws if you place them in opposite corners of the mounting flanges. The screw heads should be on the opposite side of the plate as the servo.
2. Use 4-40 × 3/4" screws and 4-40 locking nuts to attach the turntable to the turntable plate. Tighten the nuts, then back off 1/8 to 1/4 turn. The pieces should provide just enough play for floating. If things seem too rickety, tighten the nuts a bit.

Figure 13-8 shows the view from the underside of the plate; Figure 13-9 shows the top.

Figure 13-7   The base with large servo horn disc attached to the center. Note the use of nylon locking nuts. Keep the fit of the disc loose to allow the mechanism to "float" while the turntable spins.

Figure 13-8 Bottom view of the turntable plate, turntable, and base servo. Mount the servo to the turntable plate, then attach the turntable mechanism to the plate.

Figure 13-9 Top view of the turntable plate, turntable, and base servo. Nylon locking nuts (so the screws aren't held tightly) allow for play in case the shaft of the servo is not in the exact center of the turntable.

Attach the Servo C-Clamp to the Base Plate

1. Assemble a bearing on a multipurpose single servo bracket—the bearing comes in the package for the long-C bracket. The parts consist of a screw, washer, bearing, and nut. The head of the screw should be on the inside of the bracket; the washer, bearing, and nut (in that order) are on the outside of the bracket. Be sure the bearing can freely rotate after tightening the nut.

2. Temporarily mount a servo into the bracket so that the servo shaft is on the right. Place the bracket and servo on the base plate and visually align the servo shaft so that it's over the center of the plate. Mark with a pencil.

3. Remove the servo, and mark and drill two mounting holes to match any two of the small holes in the bottom of the bracket. Figure 13-10 shows an example, with screws and nuts already applied to show how the assembly will go together.

4. Using two 4-40 × 3/4" flathead screws, insert the screws from the underside of the base plate. Position a 1/8" spacer over each screw, followed by the bracket, lock washers, and nuts. The completed assembly should look like Figure 13-11.

If you don't have or can't find the 1/8" spacers you can instead use a stack of two or three #4 washers. This method is a bit harder to work with, but it gets the job done. You want a separation of about 1/8" to 3/16" between the base plate and the bottom of the servo bracket. You also want the ends of the 3/4" 4-40 screws to be flush with their nuts. If they're too long they'll interfere with the servo.

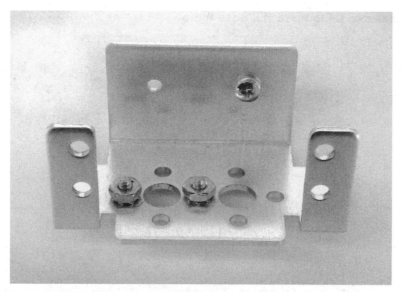

Figure 13-10 Multipurpose servo bracket, with bearing already mounted at the back, and hardware showing where the bracket will mount to the rotating base

Figure 13-11 Bracket for the shoulder servo mounted on the rotating base

Complete the Rotating Base

1. Mount the shoulder servo to the bracket you attached to the base in the previous section. You need to use only two screws for now. The shaft of the servo should face toward the right of the bracket, as shown in Figure 13-12.
2. Attach four 1" threaded standoffs to the mounting screws protruding from the bottom of the turntable plate.
3. Orient the base plate so that the shoulder servo is on the right. Orient the turntable mechanism so that its base servo is in the upper-left corner.
4. Attach the rotating base plate to the turntable plate, aligning the servo shaft with its disc.
5. Secure the round servo horn disc to the servo using the screw included with the servo.

6. Mount the base plate to the turntable mechanism using two 4-40 × 1/2" screws and locking nuts. Use the smaller of the perimeter holes in the base plate. The fasteners should be on opposite sides. Use a screwdriver and nut driver or wrench to tighten the nuts. Back off 1/8 to 1/4 turn to provide slack for the floating base.
7. Mount the completed rotating base mechanism onto the bottom plate using 4-40 × 7/16" flathead screws. Align the threaded standoffs over the screws, and tighten. See Figure 13-13 for the completed base attached to the bottom plate.

CONSTRUCTING THE UPPER ARM AND FOREARM

As a point of reference, "upper arm" is *upper* as it relates to a human arm. It's the part of our arm closest to the shoulder. On Robby, the arm is actually on the bottom—its shoulder is formed by the rotating base.

The upper arm and forearm are constructed using 1" × 9/16" U-shaped aluminum channel, cut to length as indicated in Figure 13-14. Refer to the illustration for the holes you'll need to drill.

Figure 13-12 Shoulder servo mounted into its bracket. The shaft should be oriented to the right.

Figure 13-13 Completed rotating base attached to the bottom plate. The base is secured using a set of 1" threaded standoffs.

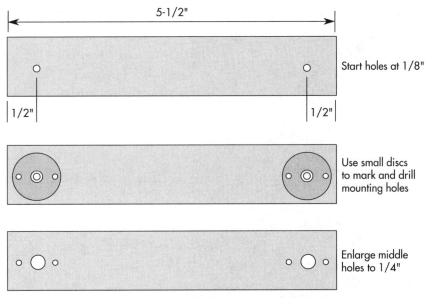

Figure 13-14 Cutting and drilling guide for the extruded aluminum U-channel used to make the upper arm. (In our case, the "upper arm" is the part connected to the shoulder joint.)

You'll need the following fasteners:

Quantity	Fastener
9	4-40 × 1/2" panhead screws
4	4-40 × 3/8" panhead screws
13	4-40 nuts
1	4-40 nylon locking nut
9	#4 washers

Build the Upper Arm

1. Drill two opposing 1/8" holes in a pair of small round servo discs (you will need to get these extra if your servos didn't come with them).
2. Temporarily attach the servo discs using a 4-40 screw and nut to the holes at the top and bottom of the upper arm piece.
3. Orient the discs so the holes in the disc are aligned vertically. Use the holes to mark for drilling using a spring-loaded punch.
4. Remove the discs, and drill the holes just marked.
5. Use two 4-40 × 3/8" panhead screws and nuts to attach a top servo disc.
6. Do likewise for the bottom servo disc, except also mount a long C-bracket at the base of the upper arm piece, as shown in Figure 13-15.

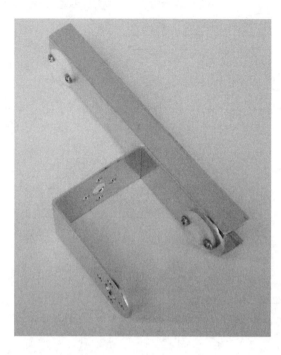

Figure 13-15 Upper arm piece, with long C-bracket attached

Build the Forearm

1. Construct two servo mounting plates using the drilling and cutting guide in Figure 13-16. Accuracy is fairly important here, or else the servos may not fit. You may also get the servo mounting plates premade if you don't want to cut them out yourself; see the parts list sources for Robby Armstrong on the ARB Support Site (see Appendix A).

2. Build the forearm using a 5-1/4" length of 1" × 9/16" aluminum U-channel. Drill holes in the channel to match those in the top of the servo mounting plates. The general layout is shown in Figure 13-17.

3. Cut a spring extension from a 1/2" wide × 1/16" thick brass bar, as shown in Figure 13-17. Drill holes approximately 1/4" apart. The exact spacing of the holes doesn't much matter; you'll use the holes to adjust the tension of the spring attached between the upper arm and shoulder. (You'll do this step after the arm is mostly finished.)

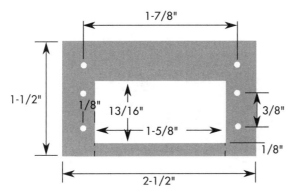

Figure 13-16 Servo mounts for use on the forearm. Hole placement and interior cut are fairly critical. The dotted lines indicate where you can cut through, if necessary, to make the hole in the center.

4. Use 4-40 x 1/2" screws and nuts and #4 washers to mount the servo plates to the forearm. The spring extension bar is mounted behind the screws on the left.
5. Secure the elbow and wrist flex servos to a mounting plate using 4-40 × 1/2" screws and nuts and #4 washers. You need only use two screws per servos, on opposite corners. The output shafts of the servos should be oriented as shown in Figure 13-18.

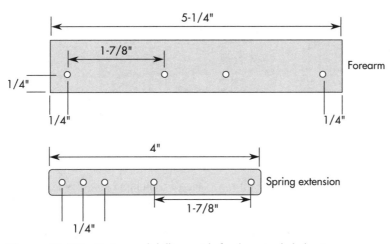

Figure 13-17 Cutting and drilling guide for the extruded aluminum U-channel used to make the forearm. The two sets of holes match those in the servo mounts shown in Figure 13-16.

Figure 13-18   Completed forearm, shown in servo brackets and servos mounted. Note the orientation of the shafts of both servos. The bar on the left is the spring extension.

BUILD THE WRIST

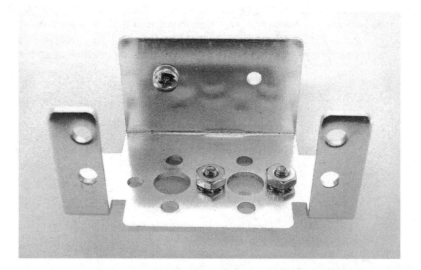

1. Assemble a bearing onto the left back side of a multipurpose single servo bracket (as before, the bearing comes in the package for the long-C bracket). Be sure the bearing can freely rotate after tightening its nut.
2. Assemble a small servo disc to the bottom of the bracket using 4-40 × 3/8" screws, nuts, and #4 tooth-lock washers.
3. Attach the servo bracket to the wrist flex servo as shown in Figure 13-19.
4. Insert and tighten the horn screw into the shaft of the wrist flex servo.
5. Secure the wrist rotation servo into the servo bracket (see Figure 13-20). Its shaft should be on the bottom.

Figure 13-19 Attach a multipurpose servo bracket (with bearing) to the wrist flex servo.

Figure 13-20 Attach the wrist flex servo into the servo bracket. The shaft of the servo should be on the bottom of the bracket.

BUILD THE GRIPPER

1. As noted previously, I'm using a commercially made gripper kit as Robby Armstrong's end-effector. This is merely a convenience for the sake of concentrating on the main aspects of building and programming the arm. You are free to use most any gripper you wish. Attach it to the arm with the L-bracket. Strive for the lightest weight gripper you can, or else it may overwhelm the other servos in the arm. Plans for the gripper I used may be found on the ARB Support Site; see Appendix A for details.
2. Attach a long C-bracket to the wrist portion of the gripper, like that in Figure 13-21. You will probably need an L-bracket (also shown). Mount a small round disc on the inside of the C-bracket as depicted.
3. Orient the gripper so that its fingers are perpendicular to the plane of the forearm. Hook the back end of the long C-bracket over the bearing on the wrist flex servo, then pop the round disc into the shaft of the wrist rotation servo, as shown in Figure 13-22.
4. Insert and tighten the horn screw into the shaft of the wrist rotation servo.

The design of the gripper used in the prototype Robby Armstrong is such that it must be only partially assembled in order to attach it to the L- and C-brackets. Once the wrist mechanics have been assembled, you can complete the construction of the gripper.

ATTACH UPPER ARM TO FOREARM

1. Orient the previously constructed upper arm so that it is at right angles to the forearm.
2. Insert and tighten the horn screw into the shaft of the elbow servo.

Figure 13-23 shows the completed upper arm and forearm, with gripper attached.

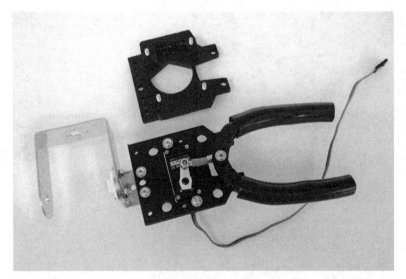

Figure 13-21 The gripper used in the prototype Robby, shown here ready to be attached to the rest of the arm. Once attached, the two halves of the gripper can be assembled.

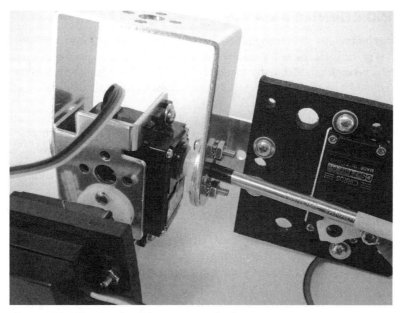

Figure 13-22 Attaching the gripper to the forearm, using a custom aluminum L-bracket (see text)

ATTACH UPPER ARM TO ROTATING BASE

1. Orient the upper arm so that it points straight up.
2. Insert and tighten the horn screw into the shaft of the shoulder servo.

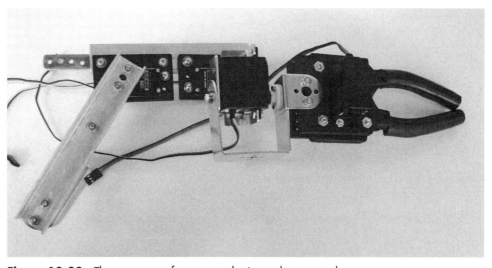

Figure 13-23 The upper arm, forearm, and gripper shown complete

ADD SPRING COUNTER BALANCE

First some background: As with many robotic revolute arms, Robby Armstrong uses a balance spring attached to the back side of the upper arm. The expansion spring provides a small amount of mechanical advantage to both the shoulder and elbow joints. The spring acts as a counterweight to allow the arm to lift more payload. Select the spring based on length and tension weight.

- The length, unexpanded, should be no more than the minimum distance between the bottom and top anchoring points. If the spring is too long, the elbow motion will be limited. Likewise, it should not be too short, or else the spring will require excessive tension when it expands. That will cause the arm servos to overwork.
- The tension of the spring is the amount of pull required to stretch it. Spring tension is largely related to the thickness of the wire and the diameter of the spring. Start with a 1/4" diameter spring with 2.5 to 3.5 pound load rating, and adjust smaller or larger from there.

You may find it beneficial to interlink two shorter springs together. This allows the two springs to "fold" onto one another, preventing the coil from jamming up the movement of the joint. To connect the springs, thread the eyelets together in the middle.

1. With the spring(s) selected, refer to Figures 13-24 and 13-25 on how to anchor the spring bottom and top, respectively.
2. For the bottom of the spring, replace the servo mounting screw with one 3/4" in length. Hold the spring in place with a #4 washer and 4-40 nylon lock nut.
3. For the top of the spring, secure in place with a 4-40 × 1/2" screw, washers, and locking nut. Start first with the center hole in the strip. As needed, shift the spring to one of the other holes to control the tension.

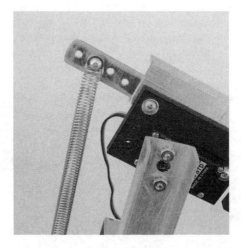

The completed mechanical portion of Robby Armstrong is shown in Figure 13-26.

Figure 13-24 Assembly detail for anchoring the bottom of the tension spring

Figure 13-25 Assembly detail for anchoring the top of the tension spring. Be sure to allow the spring to swivel at both bottom and top.

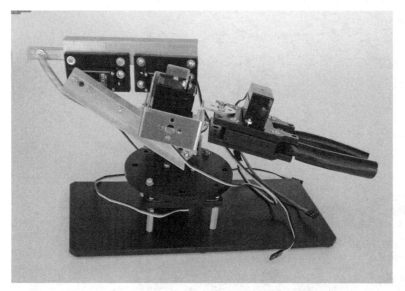

Figure 13-26 Mechanical construction of Robby shown complete

Attaching the Arduino Board

As previously noted, Robby Armstrong uses a special Arduino form factor, popularized by the DFRobot *Romeo,* where the servo headers are already integrated onto the board. This design has separate power connections for the Arduino electronics and for the servos. You can use this (or a similar) board, or you can use a standard Arduino form factor board and a servo shield. Make sure the shield supports supplying separate power to the servos.

Refer to Figure 13-27 for the placement of the Romeo-style Arduino board. Mount the board using 4-40 × 1/2" screws and nuts; add 1/8" nylon spacers to allow a small gap between the bottom of the board and the bottom plate.

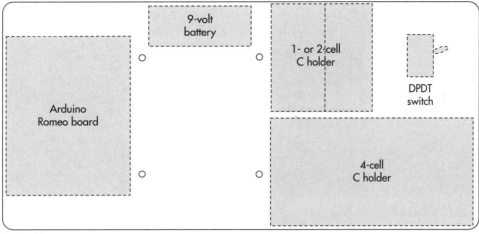

Bottom Plate

Figure 13-27   Suggested component placement on the bottom plate: Arduino Romeo board (see text), battery holders, and power switch

Wiring for Power

Robby has a prodigious appetite for power. It has two power sources, both of which are batteries:

- A 9-volt battery provides power for the Arduino and any connected electronics (keypad, LCD, optional ultrasonic sensor).
- A set of five C-size rechargeable NiMH batteries provide juice for the servos. Each C cell outputs a nominal 1.2 volts; the set of five generates 6 volts. (If you are using servos that can tolerate 7.2 volts—which some are able to do—you can use a set of six batteries. This is what I did for the prototype Robby; the higher voltage provides more torque.) The batteries are held in two holders: one for four cells, and the other for one or two cells. The holders are wired in series, shown in Figure 13-28.

If using a custom-built board for connecting the servos to the Arduino, be sure to connect the grounds between the Arduino and servos. Common ground is already provided in boards like the Romeo and in servo shields.

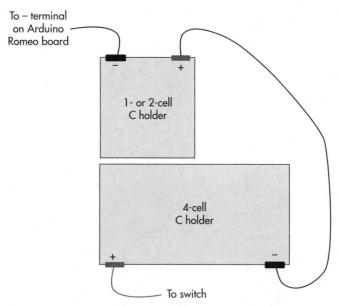

To – terminal on Arduino Romeo board

1- or 2-cell C holder

4-cell C holder

To switch

Figure 13-28 Wire the two C-cell holders in series to increase the voltage. Using a four-cell holder and a two-cell holder provides 7.2V (with rechargeable NiMH batteries). I used servos in the prototype Robby rated for 7.2V; if your servos are rated only for 6V, use a four- and one-cell holder (6V).

A double-pole, double-throw (DPDT) switch controls the power from the batteries (see Figure 13-29). In Position 1, only the 9-volt circuit is connected. This allows you to program the Arduino and check electronic control. In Position 2, both power supplies are connected to their respective circuits. This is the regular operating mode when you wish to work the arm.

For ease of use mount the switch onto the baseplate. On my prototype I drilled holes in the baseplate and passed the wire from the batteries and switch through them.

The power wiring itself is on the underside of the baseplate. A set of six rubber feet act as pads for the baseplate, and provide clearance for the wiring and any protruding hardware. Figure 13-30 shows a portion of the wired underside.

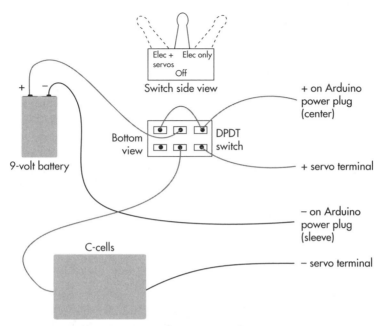

Figure 13-29 Wiring for Robby's main switch. Use a DPDT switch with a center off. In one direction, the switch powers just the Arduino and electronics. In the other direction, the switch powers both Arduino and servos.

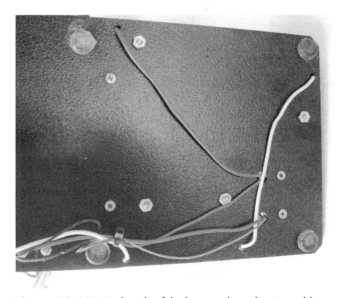

Figure 13-30 Underside of the bottom plate, showing rubber feet. The feet act as spacers for any hardware or wiring underneath.

Connecting the Control Circuits to the Arduino

Robby sports an LCD panel and 12-key keypad that communicate with the Arduino via I2C serial. The I2C link uses Arduino pins A4 and A5, plus needs 5V power and ground. The Romeo board provides several four-pin headers to make it easier to connect to I2C devices. Likewise, the arm uses two 5-position switches to control joint movements.

Figure 13-31 shows how the servos, LCD/keypad, and switches connect to the Romeo Arduino board.

The Romeo does not include the usual pull-up resistors often used when interfacing I2C devices. The pull-ups connect between 5V and the A4 and A5 pins on the Arduino. I found everything worked fine without them, but if you observe erratic behavior, you might want to add the resistors. Typical values are between 1.8 KΩ and 10 KΩ, with many folks recommending values in the lower range. (I usually start with 3.3 KΩ and see how that fares.)

Note that you need only one resistor for each I2C line (pins A4 and A5), not for every I2C device you have attached to the Arduino.

Figure 13-32 shows a suggested layout for the LCD, keypad, and switches on a separate controller board. As depicted in Figure 13-33, I've mounted the switches on a half-size solderless breadboard, then routed the wiring from the breadboard to the Arduino.

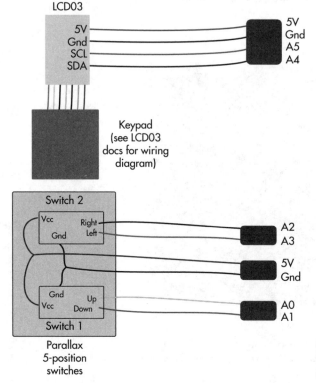

Figure 13-31 Wiring diagram for the LCD, keypad, and five-position switches

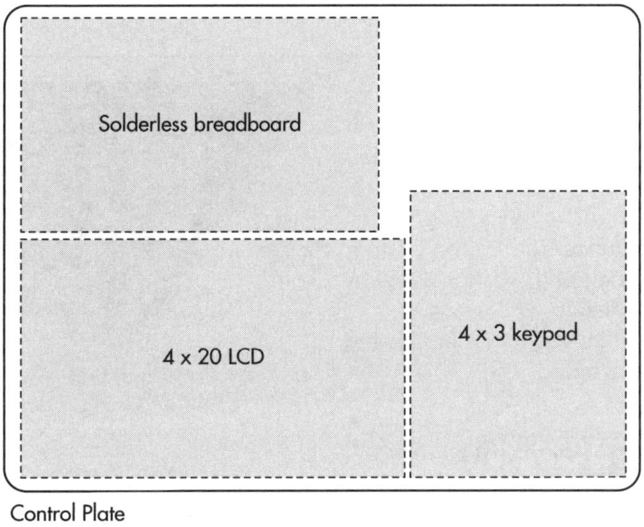

Control Plate

Figure 13-32   Suggested component placement on the control plate: LCD, solderless breadboard (for the five-position switches), and 12-key keypad

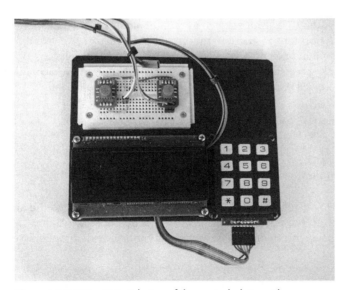

Figure 13-33   Pictorial view of the control plate, with components mounted. Note the LCD specified communicates with the Arduino via I2C, and it has an integrated adapter to connect to a keypad (up to 4×4 matrix).

Adding Rubber Feet

For a more professional look and feel add a set of rubber pad "feet" to the underside of both the arm bottom base and the control board. The feet should be at least 1/8" thick.

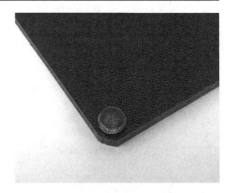

- Apply six feet to the underside of the bottom base; one in each corner, and additional feet in the center along the sides.
- Apply four feet, one in each corner, to the underside of the control board.

Programming Robby

With construction complete you're now ready to calibrate and use Robby Armstrong.

CALIBRATING THE HOME AND MIN/MAX POSITIONS

Refer to the sketch *Robby_Calibrate* for setting the "home" positions of each of the six servos. These home positions may not—and likely are not—the same as the 1500 μs neutral or center position of the servos. The home position is where the servos already start from whenever you apply power to Robby.

You will also use the *Robby_Calibrate* sketch to set the minimum and maximum positions for the six servos. These positions restrict the movement of the servos, for example, to prevent part of Robby from hitting against some other part. As you use the sketch, jot down the values you obtain for each servo so you can modify Robby's main programs accordingly. The values you determine here will be used in other sketches for operating Robby.

 To save page space, the Robby_Calibrate sketch is provided on the ARB Support Site. See Appendix A for details.

To use the *Robby_Calibrate* sketch:

1. Position Robby's joints following their positions when you constructed the arm.
2. Upload the sketch to the Arduino, then open the Serial Monitor window. Make sure it's set to 9600 baud communications, and select the line-ending option to Carriage Return.
3. Apply power to the servos, then reset the Arduino. The servos will suddenly (and swiftly!) move to their neutral positions.

In the Serial Monitor window enter a servo move command in the following format:

```
s.n.xxx
```

where *n* is the number of the servo you wish to move, and *xxx* is its position, in 0 to 180 degrees (90 degrees is center). For example, to move servo 2 (elbow) to 100 degrees, enter

```
s.2.100
```

then press the ENTER key on your computer.

The servo numbers and their corresponding joints are:

Servo #	Joint
0	Base
1	Shoulder
2	Elbow
3	Wrist flex
4	Wrist rotation
5	Gripper

 Manually moving the position of servos is called back-driving. Servos can become damaged if they are back-driven too quickly. Slowly rotate the servos to their desired position. Avoid rapidly spinning the servo, or else its internal gearing may become damaged. If that happens you'll need to either replace the gears or replace the servo.

For each servo:

- Establish a startup home position for the servo. Use a "stance" like that shown in Figure 13-34 as the startup position for the joints. Whenever you want to activate Robby, manually move the joints to this position, then power up the arm.
- Establish a minimum and maximum position; allowable values are between these two extremes. For the purposes of Robby's programming—and somewhat counterintuitively— the *minimum* position is the greater value toward 180. The maximum position is the lower value toward 0.

Once you have the home, minimum, and maximum positions identified, format them in a code structure like this:

```
ServoPos servosPos[] = {
  { 60, 160, 0, 0, 0 },
  { 130, 180, 120, 0, 0 },
  { 110, 150, 60, 0, 0 },
  { 20, 160, 5, 0, 0 },
  { 110, 180, 10, 0, 0 },
  { 95, 160, 20, 0, 0 }
};
```

Figure 13-34   Robby in its home stance. The servos are set to their "home" position for startup. (Note the ultrasonic sensor on the gripper. This is an optional add-on to Robby to enable it to know if there's something within grasp, or to perform 3D mapping. Feel free to add one if you'd like to do further development.)

Each line of code represents a servo, from servo 0 to servo 5. The first column in each line is the startup neutral home position, followed by the minimum and then maximum positions. (The two sets of 0 values on each line are variables used internally by the Robby Armstrong sketches; just leave them at 0.)

For instance, in the third line (the elbow servo) you see

```
{ 110, 150, 60, 0, 0 },
```

which indicates a startup neutral (home) position of 110, minimum of 150, and maximum of 60. These were values I determined for my prototype Robby; your position values will likely be different.

USING THE SERIAL MONITOR FOR CONTROLLING THE ARM

The most straightforward method of controlling Robby Armstrong is to use the Serial Monitor window to send commands to the Arduino, just as we did with *Servo_Calibrate*. The code in the *RobbySerialMonitor* sketch allows you to enter a simple command sequence that specifies the servo to move, and where you want to move it to. As with the calibration sketch, servos are numbered 0 through 5, and positions are set using angular degrees, from 0 to 180.

Of critical importance is the *typedef structure,* which—as you learned earlier—sets the home position, minimum value, and maximum value of the servos. You were introduced to this structure in the previous section where you made changes to the values for each servo. Update the code in *RobbySerialMonitor* to use your own startup neutral (home), minimum, and maximum values.

The sketch works by establishing an array (named *servoPos*) that stores six separate copies of the structure: *servoPos[0]* for servo 0, *servoPos[1]* for servo 1, and so on. Each element of the array holds values for one servo. The *minPos* and *maxPos* values of the structure disallow you from adjusting the position of each joint beyond its allowed extents. The neutral value stores the startup home position, which is where you want the servos to move to whenever Robby is first powered on.

RobbySerialMonitor

```
// Format: s.[servonum].[pos]. Ex: s.1.145 (servo #1 to position 145)
// Be sure to set Serial Monitor to Carriage Return

#include <Messenger.h>              // Include Messenger object library
#include <Servo.h>

const int maxServos = 6;            // Total number of servos
const int offset = 4;              // Begin at D4

// These are for reference and further expansion by you
const int base =         0;
const int shoulder =     1;
const int elbow =        2;
const int wristflex =    3;
const int wristrot =     4;
const int gripper =      5;

typedef struct{                    // Structure holds values for each servo
  byte neutral;
  byte minPos;
  byte maxPos;
  byte delaySpeed;                 // Not used in this version
  byte curPos;
} ServoPos;

ServoPos servosPos[] = {           // Array of ServoPos structures
  { 60, 160, 0, 0, 0 },
  { 130, 180, 120, 0, 0 },
  { 110, 150, 60, 0, 0 },
  { 20, 160, 5, 0, 0 },
  { 110, 180, 10, 0, 0 },
  { 95, 160, 20, 0, 0 }
};

Servo servos[maxServos];                    // Array of servo objects
Messenger message = Messenger('.');         // Messenger object constructor

void setup() {
  Serial.begin(9600);
  for(int i=0; i<maxServos; i++) {          // 0 to 5
    servos[i].write(servosPos[i].neutral);  // Preset servos to neutral
    servos[i].attach(i+offset);             // Attach servo, starting at D4
    delay(1000);
  }
  message.attach(messageCompleted);   // Attach Messenger to function
}
```

```
void loop() {
  while (Serial.available()) {          // If bytes in Serial port
    message.process(Serial.read());     // Read message in Serial port
  }
  delay(100);
}

void messageCompleted() {               // Process if message starts with 's'
  if (message.checkString("s"))
    writeServo();
}

void writeServo() {
  int servoNum = message.readInt();     // Get servo number
  if(servoNum >=0 && servoNum <= maxServos) {
    byte t = message.readInt();         // Get servo position
    // Constrain values to max and min
    servosPos[servoNum].curPos = constrain(t, servosPos[servoNum].maxPos,
      servosPos[servoNum].minPos);
    servos[servoNum].write(servosPos[servoNum].curPos);   // Move to position
    Serial.print("Servo: ");            // Print servo number and position
    Serial.print(servoNum, DEC);
    Serial.print(", Position: ");
    Serial.println(servosPos[servoNum].curPos, DEC);
  }
}
```

The *RobbySerialMonitor* sketch uses a third-party library, *Messenger*, to intercept text strings from (among other things) the Serial Monitor window. The command string is in the format s.[servonum].[pos]

where

- *s* is the command prefix
- *[servonum]* is the number of the servo to move (from 0 to 5)
- *[pos]* is the position you want to move the servo to (0 to 180)

For example:

```
s.3.100
```

means move servo 3 to position 100.

 Press the ENTER key after typing a command string. Be sure to set the line-ending option in the Serial Monitor window to Carriage Return.

Also of importance in this sketch is the array created using the *ServoPos* typedef structure. Here it is again, calibrated for my prototype of Robby:

```
ServoPos servosPos[] = {
  { 60, 160, 0, 0, 0 },
  { 130, 180, 120, 0, 0 },
  { 110, 150, 60, 0, 0 },
  { 20, 160, 5, 0, 0 },
  { 110, 180, 10, 0, 0 },
  { 95, 160, 20, 0, 0 }
};
```

As you learned in the previous section, you will likely need to adjust these to better match your particular hardware. The first three columns of each array contain the neutral (startup home), minimum, and maximum position:

Column 1	Column 2	Column 3
Neutral position	Minimum position	Maximum position

 I touched upon this earlier, but in case you're interested, the other two columns of each array hold the delay speed and current position, respectively. Delay speed is not used in this sketch, and the current position column—named *curPos*—is updated during the running of the sketch.

USING JOYSTICK AND KEYPAD CONTROLLERS

You can operate Robby Armstrong independently from your computer using the "teaching pendant" controller. There are all kinds of pendant controllers used for robotics, such as game-style joysticks, wearable sensors, gyros, accelerometers, and other forms of inertial monitors.

Recall that the pendant for Robby is composed of three parts:

- Two 5-position switches provide directional control to all joints in the arm. One switch changes the joint you wish to control; the other adjusts the position up or down. I'm using two 5-position switches out of convenience; you can construct the circuit with just one. Each switch has momentary spring-loaded contacts for up, down, right, and left movement, plus center detent.
- A 4-line-by-20-character LCD display provides feedback as you move the arm joints. I'm using an LCD that connects to the Arduino via I2C.

Many LCDs use standardized interface controllers, making it much easier to work with them. The controller is the electronics built into the LCD that provides the communications gateway. For text displays, the most common controller is the Hitachi HD44780. There are three primary ways to interface to this controller, and thereby change the text on the LCD. The methods are parallel, asynchronous serial, and I2C serial.

Parallel interfacing involves directly connecting separate I/O pins from the Arduino to the LCD controller. Text-based LCDs require a minimum of seven I/O pins.

Asynchronous serial interfacing involves connecting two or three I/O pins. The Arduino communicates with the LCD via eight-bit serial commands. Additional electronics on the LCD convert these commands to the parallel interface used on the panel.

I2C serial interfacing also uses additional electronics to connect Arduino to LCD. I2C is a bidirectional synchronous serial protocol that uses two I/O pins: one pin carries the data, the other a clock pulse that provides the synchronization.

Of these types, asynchronous serial and I2C serial are the easiest to interface to the Arduino, as they require as few as one pin connection (in addition to power and ground). For the prototype Robby Armstrong I'm using the LCD03 from Devantech, which combines a 4-line-by-20-character backlit LCD panel with a built-in 4×4 keypad matrix decoder. The keypad selected for Robby contains just 12 keys; these also work with the LCD03.

Programming Robby for Interactive Control

The main programming code for Robby Armstrong is in a sketch named *Robby_Control*. It's rather lengthy and fairly involved, so instead of listing it in its entirety here I'll break it out into main chunks, then detail what these chunks do. You'll want to fetch the full sketch from the ARB Support Site provided in Appendix A.

```
#include <Servo.h>
#include <Wire.h>
#include <TimedAction.h>
#include <LCDi2cR.h>
```

The sketch uses four libraries; the first two (*Servo* and *Wire*) come with the Arduino IDE, and the other two (*TimedAction* and *LCDi2cR*) are from third-party sources, and must be added to the Arduino sketchbook libraries folder prior to compiling. The ARB Support Site provides download links for these libraries.

```
const int base =        0;
const int shoulder =    1;
const int elbow =       2;
const int wristflex =   3;
const int wristrot =    4;
const int gripper =     5;
```

The six servos are given names to make them easier to work with. Each servo is assigned a numeric value, starting at zero. Added with an offset value (defined later in the sketch) these numbers are used to define the Arduino pins each servo is connected to.

```
const int stdDelay =    20;     // Ms between each servo movement
const int maxServos =   6;      // Max number of servos
const int centerPos =   90;     // Default center position
int keyDelay = 100;             // Max delay (ms) between each key
int buttonDelay = 50;           // Min delay between each key
int thisServo = base;           // Default servo to start
unsigned long key_millis = 0;   // Wait timers
unsigned long button_millis = 0;
```

Setup variables provide some customization to the sketch. Of particular importance is the *stdDelay* variable. It sets the time, in milliseconds, the servos wait between incremental movement. This delay allows the servos to move more slowly; if the servos rotate too quickly, the arm reacts in a nervous, jerky fashion.

```
typedef struct{
  byte neutral;
  byte minPos;
  byte maxPos;
  byte delaySpeed;
  byte curPos;
} ServoPos;
ServoPos servosPos[] = {
  { 60, 160, 0, stdDelay, 0 },
  { 130, 180, 120, stdDelay, 0 },
  { 110, 150, 60, stdDelay, 0 },
  { 20, 160, 5, stdDelay, 0 },
  { 110, 180, 10, 10, 0 },
  { 95, 160, 20, 5, 0 }
};
```

Various setting values for the servos are stored in a structure, which is basically a collection of variables that are prefixed by a standard name. The structure itself is just a framework; to use it you must define a variable based on the structure. This is done by defining a variable with the name of the structure, which is *ServoPos*. For example:

```
ServoPos myServo;
```

creates a variable called *myServo* that is based on the *ServoPos* structure. Robby uses six servos, and while each servo could have its own variable, to make things easier the sketch uses an array called *servoPos[]*—the [] denotes this is an array. As we've already discussed, the array holds six elements, one element per servo.

Notice the structure name is *ServoPos* and the variable defined from the structure is *servoPos*. The names are similar, but not identical. The sketch treats them as distinct. The names follow an internal naming convention, where structures are camelcase with the first letter capitalized, whereas with the variable the first letter is lowercase.

Of course, you are free to use your own naming convention. You may find it helpful to give the variables based on structures a similar name as the structure itself, so you can keep track of which variables belong to what structures.

Default values are stored for each array element at the time the *servoPos* array is defined. The meaning of each value is defined in the *ServoPos* structure, and each value is defined in order:

- *neutral* is the startup home position of the servo. The servo returns to this position whenever power is applied or the sketch is restarted.
- *minPos* is the minimum position of the servo allowed. This is set to prevent any joint from going beyond the mechanical limit.
- *maxPos* is the maximum position of the servo allowed.
- *delaySpeed* is the time delay between each incremental move of the servo. Most of the joints use a default delay set using the stdDelay variable, defined earlier in the sketch. But any value can be used here, which is the case with the gripper, so that the fingers open and close more slowly.
- *curPos* is the current position of the servo. This value is set by the sketch as Robby's joints are moved.

```
LCDi2cR lcd = LCDi2cR(4,20,0x63,0); // Set up 4x20 LCD for I2C

Servo servos[maxServos];
int destServoPos[maxServos];
int currentServoPos[maxServos];

TimedAction servoMove[maxServos] = TimedAction(100, doServoFunc);
TimedAction keys = TimedAction(10, keypadFunc);
```

Several objects are set up to define the LCD interface, the six servos, and a couple of background timers. The third-party *LCDi2cR* library handles both an LCD and 12/16-key matrix keypad (Robby uses a 12-key keypad), both through the I2C interface on the Arduino. I've set the LCD object for a 20-character-by-4-line display; if yours is different, modify the first two parameters accordingly.

Notice that Robby's six servos are defined as objects in an array named *servos[]*. This allows the servos to be easily referenced by number.

Robby's sketch doesn't rely on the Arduino *delay* statement to move servos. This allows multiple servos to move—at any speed—at the same time. Instead, it uses a background timer using the third-party *TimedAction* object library. This object lets you trigger an action at preset intervals.

There are actually seven *TimedAction* timers: six control the servos, and a seventh is for checking the keypad for any button presses. Knowing how these timers work is critical to understanding the sketch, so here's some more detail.

Each object defined using the *TimedAction* library requires two parameters, a timeout interval (in milliseconds), and a function within the sketch that is called each time the timeout interval elapses. In the case of the six *servoMove TimedAction objects* (one object for each servo), the interval is once every 100 milliseconds. The same function, *doServoFunc,* is called regardless of the object. But note that each loop through the function is distinct.

```
void setup() {
  delay(200);
  Wire.begin();              // Start I2C commn
  lcd.init();                // Initialize LCD (uses I2C)
  delay(500);
  lcd.on();              // Turn on LCD backlight
  setupDisplay();

  for(int i=0; i<maxServos; i++) {
    servos[i].write(servosPos[i].neutral);   // Set neutral
    servosPos[i].curPos = servosPos[i].neutral; // Set curPos to neutral
    servos[i].attach(i+4);                    // Attach servo to pin
    destServoPos[i] = centerPos;
    currentServoPos[i] = centerPos;
    servoMove[i].disable();                   // Disable timer for servo
  }
}
```

The setup function starts the I2C communication link (via the Arduino Wire object) to the LCD. It then goes through each of the six servos and sets them up for operation. The current position for each servo is set to its neutral, attached to the appropriate pin (pins D4 through D9). The timer that operates the servo movement (one timer per servo) is then disabled.

Recall that the *servoPos* array is defined from the *ServoPos* structure. Note the dot operator between the array name and the name of the variable within the structure. For example, to set the current position of the servo, the line

```
servosPos[n].curPos
```

references one of the six servos (defined as *n,* a number from 0 to 5), and the *curPos* element within the *ServoPos* structure. The dot operator is the most common way of working with structure elements.

(There are some instances where you use the -> "arrow" operator instead of the dot operator. None of the sketches in this book require this technique, but you may encounter it in other sketches you work with. I mention it only so you can be aware of it in case you encounter it someday.)

```
void loop() {
  for(int x=0; x<maxServos; x++) {
    curServo = x;
    servoMove[x].check();
  }
  keys.check();
  navSwitchFunc();
}
```

The main loop of the sketch iterates through the servos, triggering the *TimedAction* software timer associated with each servo—remember, six timers for six servos. The timer moves the servo as needed. Though not heavily used in this demo sketch, the construction of the loop allows the servos to move in slow motion concurrently.

Also in the loop is the check-for button pressed on the 12-key keypad (also via a *TimedAction* timer). This function is also provided for using a software timer. This timer causes the keypad to be examined every 10 milliseconds.

Again note there aren't any Arduino *delay* statements in the main loop. This is very important.

```
void Move(int servoNum, int servoPosition, int delayTime) {
  sMove[0] = servoNum;
  sMove[1] = servoPosition;
  sMove[2] = delayTime;
  writeServo();
}
```

The *Move* function sets up servo movement by first filling in an array that (out of convenience for future expansion) is shared across some functions. Servo move instructions entail three parts:

- The number of the servo to move, 0 to 5;
- The final servo position; and
- The delay period between each incremental servo move.

```
void writeServo() {
  int servoNum = sMove[0];
  if(servoNum >=0 && servoNum <= maxServos) {
    destServoPos[servoNum] = sMove[1];
    servoMove[servoNum].enable();
    servoMove[servoNum].setInterval(sMove[2]);
  }
}
```

The *writeServo* function moves a given servo, using the data points previously slugged into the *sMove* array. An *if* test first determines if the servo is available, then defines the destination position. The *TimedAction* timer for the servo is enabled, and its interval timeout is set.

```
void doServoFunc() {
  int x = curServo;                // Don't use global variable here

  if(destServoPos[x] == currentServoPos[x])
    servoMove[x].disable();
```

```
  if(destServoPos[x] > currentServoPos[x])
    currentServoPos[x]++;
  else
    currentServoPos[x]--;

  servosPos[x].curPos = constrain(currentServoPos[x], servosPos[x].maxPos,
    servosPos[x].minPos);
  currentServoPos[x] = servosPos[x].curPos;
  servos[x].write(currentServoPos[x]);
  jointPos(x, currentServoPos[x]);
}
```

The *doServoFunc* is something of the nerve center of the Robby sketch. The function incrementally moves a servo from its current position to its final (destination) position. It does this move one degree at a time. The function is skipped if the servo is already at its destination point.

```
void navSwitchFunc() {
  if (millis() > button_millis + buttonDelay) {
    button_millis = millis();
    if(digitalRead(A0) == LOW) {
      thisServo--;
      thisServo = constrain(thisServo, 0, 5);
      jointPos(thisServo, servosPos[thisServo].curPos);
      delay(200);
    }

    if(digitalRead(A1) == LOW) {
      thisServo++;
      thisServo = constrain(thisServo, 0, 5);
      jointPos(thisServo, servosPos[thisServo].curPos);
      delay(200);
    }

    if(digitalRead(A2) == LOW) {
      byte t = thisServo;
      servosPos[t].curPos--;
      servosPos[t].curPos = constrain(servosPos[t].curPos,
        servosPos[t].maxPos, servosPos[t].minPos);
      jointPos(t, servosPos[t].curPos);
    }

    if(digitalRead(A3) == LOW) {
      byte t = thisServo;
      servosPos[t].curPos++;
      servosPos[t].curPos = constrain(servosPos[t].curPos,
        servosPos[t].maxPos, servosPos[t].minPos);
      jointPos(t, servosPos[t].curPos);
    }
  }
}
```

Robby has two means of control, both demonstrated in the sketch. There's the more elaborate 12-key keypad, described next, or a pair of five-position switches. One switch selects the servo to move; the other switch changes the servo position. *navSwitchFunc* reads the value of the switches (on Arduino pins A0 to A3) and runs the associated code for that switch movement. Note that on the Parallax five-position switch called out for use in Robby, the switch output is normally HIGH when the contacts are open. The output goes LOW when the switch makes contact.

You can combine the Up/Down and Left/Right actions and use just one switch. I used two 5-position switches instead of one for ease of control. It's possible to move a switch both up *and* right, for example. Having two switches reduces the chance of accidentally selecting a mode you don't want. Having two switches also allows for some future expansion.

```
void keypadFunc() {
  byte keypad = lcd.keypad();
  lcd.command(0);                        // Send null to start
  if (keypad !=0) {
    if (millis() > key_millis + keyDelay) {
      keypress(keypad);
      key_millis = millis();
    }
  }
}
void keypress (byte keypad) {
  byte t;
  lcd.setCursor(0, 5);
  switch(keypad) {
    case 1:
      lcd.print("1");
      t = base;
      servosPos[t].curPos--;
      break;
  // cases 2 through 14
    case 15:
      lcd.print("#");
      t = gripper;
      servosPos[t].curPos++;
      break;
  }

  servosPos[t].curPos = constrain(servosPos[t].curPos,
    servosPos[t].maxPos, servosPos[t].minPos);
  jointPos(t, servosPos[t].curPos);
}
```

The *keypadFunc* and *keypress* functions read the buttons on the 12-key keypad. Each button is associated with a specific servo and motion. For example, pressing the 1 button moves the base by decrementing the current position of the base servo.

As with many libraries for dealing with multi-key keypads, the one used in Robby is set up for a 4x4 matrix, allowing for up to 16 buttons. We're only using 12 for Robby Armstrong. Not all of the button keys are referenced in the code. The ones that are missing are for a 16-key keypad.

```
void jointPos(byte t, byte pos) {
  lcd.setCursor(1, 7);
  switch(t) {
    case base:                  // 1, 4
```

```
        lcd.print("bse");
        break;
// remaining cases perform same function as above
    }

    lcd.setCursor(1, 17);
    lcd.print("   ");
    lcd.setCursor(1, 17);
    lcd.print(pos, DEC);
    servos[t].write(pos);
}
void setupDisplay() {
    lcd.clear();
    lcd.blink_off();
    lcd.home();                      // Cursor to home position (0,0)
    lcd.print("key: ");
    lcd.setCursor(1, 0);
    lcd.print("Joint:      ");
    lcd.setCursor(1, 12);
    lcd.print("Pos: ");
}
```

The *jointPos* function is part of the LCD display aspect of Robby. Motion control via the five-position switches or 12-key keypad is shown on the attached LCD, which as you'll recall from earlier is commanded via the third-party *LCDi2cR* library. This library supports the 4 × 20 LCD displays made and sold by Devantech. This LCD has a connector on it for attaching to a 12- or 16-key keypad. Through this library the LCD is commanded to display the servo name and its position.

The *setupDisplay* function is called at the beginning of the sketch, and sets up the LCD by clearing it (in case there's a remnant of text from before), turns off the cursor blinking mode, and inserts initializing text.

Operating Robby Armstrong

You may operate Robby's six servos using either joysticks or keypad.

OPERATING THE JOYSTICKS

The joysticks select and incrementally move the servos, respectively:

- Move stick #1 up or down to select the servo you wish to move.
- Move stick #2 to the left or right to change the servo position.

The currently selected servo and its position are shown on the LCD.

 Robby's control sketch prevents you from moving a servo beyond its minimum and maximum extends. If you try to position a servo and nothing seems to happen, check that the servo has not stopped at a min/max extent.

OPERATING THE KEYPAD

The keypad provides an alternative, and quicker, method of selecting and incrementally moving servos. Each button is dedicated to moving a servo up or down:

Buttons	Controls Joint
1 and 4	Base
2 and 5	Shoulder
3 and 6	Elbow
7 and *	Wrist flex
8 and 0	Wrist rotation
9 and #	Gripper

As with joystick control, the currently selected servo and its position are displayed on the LCD.

Part 4

Appendixes

ARB Online Support

Books have only so many pages, so to augment your reading experience, be sure to visit the ARB Online Support site at *www.robotoid.com*.

You'll Find...

Book updates   Changes, corrections, enhancements, alternative approaches.
Program listings for all the projects   Download the program files for immediate use.
Expanded parts source for all the projects   Including online sources, with part numbers and direct links to the products (when available). Just click and you're there.
Plan templates drawn to scale   Numerous projects and parts. Just download the PDF, and print.
Video tutorials   On everything from using shop tools to programming your bot.

Plus, free articles, expanded projects, news, reviews, and more.

Sources for Special Parts and Web Sites

Mentioned now and then in this book are special parts and Web sites for improving your robot building experience. You can find some of these sites in Appendix B, "Parts Connection," but you'll find even more on the ARB Online Support site. They're categorized, with a brief summary of the most important sites on the list.

Parts Connection

O n these pages you'll find a selected list of online sources for robots and robot building parts. I haven't listed everyone, just a handful of the best, and those that have been around for a while.

FYI Web sites come and go. New and updated sources may be found at the ARB Online Support site. See Appendix A for details.

Robotics

Robot kits, sensors, servos, wheels, parts.

Acroname—www.acroname.com
Arrick Robotics—www.robotics.com
Budget Robotics—www.budgetrobotics.com
CrustCrawler—www.crustcrawler.com
Fingertech Robotics—www.fingertechrobotics.com
Lynxmotion—www.lynxmotion.com
Machine Science—www.machinescience.com
Mark III Robot Store—www.junun.org/MarkIII/
Mr. Robot—www.mrrobot.com
Parallax—www.parallax.com

Pitsco—www.pitsco.com
Pololu—www.pololu.com
Robot MarketPlace—www.robotmarketplace.com
Robotis—www.robotis.com
Robot Store (HK)—www.robotstorehk.com
RobotShop—www.robotshop.com
Solarbotics—www.solarbotics.com
TrossenRobotics—www.trossenrobotics.com
Vex Robotics—www.vexrobotics.com
Zagros Robotics—www.zagrosrobotics.com

Electronics

Components, modules, microcontrollers. May include both new and surplus.

Allied Electronics—www.alliedelec.com
Arrow Electronics—www.arrow.com
Circuit Specialists—www.web-tronics.com
Devantech Ltd.—www.robot-electronics.co.uk
Dick Smith Electronics—dicksmith.com.au
Digi-Key—www.digikey.com
Electronix Express—www.elexp.com
Farnell—www.farnell.com
Future Electronics—www.futureelectronics.com
Hobby Engineering—www.hobbyengineering.com
HVW Tech—www.hvwtech.com
Images SI—www.imagesco.com
Jameco Electronics—www.jameco.com
Maplin Electronics—www.maplin.co.uk
Marlin P. Jones & Assoc.—www.mpja.com
MCM Electronics—www.mcmelectronics.com
Mouser Electronics—www.mouser.com
Newark Electronics—www.newark.com
Nu Horizons Electronics Corp.—www.nuhorizons.com
Parts Express—www.partsexpress.com
RadioShack—www.radioshack.com
SparkFun Electronics—www.sparkfun.com

Hobby

Servo motors and accessories, radio control sets, parts for model airplanes and cars. Many of the sources listed under "Robotics" and "Electronics" also carry these parts.

Hobby King—www.hobbyking.com
Hobby Lobby—www.hobbylobby.com
Hobby People—www.hobbypeople.net

Horizon Hobby—www.horizonhobby.com
Servo City—www.servocity.com
Tower Hobbies—www.towerhobbies.com

Specialty Sources

The following sources provide unique products that are useful in the design and construction of amateur robots.

BRACKETS: METAL AND PLASTIC

Amazon Supply—www.amazonsupply.com
Aubuchon Hardware—www.hardwarestore.com
Budget Robotics—www.budgetrobotics.com
CornerHardware.com—www.cornerhardware.com
Rockler Woodworking and Hardware—www.rockler.com

COMPETITIONS AND ROBOTICS CAMPS

Atlanta Hobby Robot Club—www.botlanta.org
Battlebots—www.battlebots.com
Botball—www.botball.org
Carnegie Mellon Robotics Club—www.roboticsclub.org
Central Illinois Robotics Club—circ.mtco.com
Chicago Area Robotics Group—www.chibots.org
Dallas Personal Robotics Group (DPRG)—www.dprg.org
Portland Area Robotics Society—www.portlandrobotics.org
Robocup—www.robocup.org
Robotics Society of Southern California—www.rssc.org
Seattle Robotics Society—www.seattlerobotics.org
Titan Robotics Club—www.titanrobotics.net
US FIRST—www.usfirst.org

ELECTRONIC KITS

Carls Electronics—www.electronickits.com
Chaney Electronics—www.chaneyelectronics.com
Quasar Electronics—www.quasarelectronics.com
Ramsey Electronics—www.ramseyelectronics.com

FORUMS (NON-PRODUCT SPECIFIC)

Google Groups—groups.google.com
Let's Make Robots—www.letsmakerobots.com
Lugnet—news.lugnet.com/robotics
Nuts & Volts/SERVO Bulletin Board—www.servomagazine.com/forum
Robotics Club of Yahoo (TRCY)—groups.yahoo.com/group/theroboticsclub

GEARS AND MECHANICAL

Amazon Supply—www.amazonsupply.com
Berg—www.wmberg.com
Serv-o-Link—www.servolink.com
Servo City—www.servocity.com
Stock Drive Products—www.sdp-si.com

PLASTICS AND SUBSTRATES

Budget Robotics—www.budgetrobotics.com
R & J Sign Supply—www.rjsign.com
Regal Plastics—www.regal-plastics.com
Ridout Plastics—www.ecomplastics.com
TAP Plastics—www.tapplastics.com

SURPLUS

Products may include new, used, and surplus, both mechanical and electronic.

A-2-Z Solutions—www.a2z-solutions.com
All Electronics—www.allelectronics.com
Alltronics—www.alltronics.com
American Science & Surplus—www.sciplus.com
Apex Jr.—www.apexjr.com
B.G. Micro—www.bgmicro.com
Burden Surplus Sales—www.surpluscenter.com
C & H Surplus—candhsurplus.com
CTR Surplus—www.ctrsurplus.com
Electro Mavin—www.mavin.com
Electronic Goldmine—www.goldmine-elec.com
Electronic Surplus—www.electronicsurplus.com
Excess Solutions—www.excess-solutions.com
Fair Radio Sales—www.fairradio.com
HSC Electronic Supply (Halted)—www.halted.com
Hosfelt Electronics—www.hosfelt.com
Marlin P. Jones & Assoc.—www.mpja.com
Mendelson Electronics Company—www.meci.com
Skycraft Parts & Surplus—www.skycraftsurplus.com
Surplus Sales of Nebraska—www.surplussales.com
Timeline—www.timeline-inc.com
Unicorn Electronics—www.unicornelectronics.com
Weird Stuff Warehouse—www.weirdstuff.com

How to Find Electronic Parts in a Big Catalog

So you want to locate that 10 µF non-polarized aluminum electrolytic capacitor you need for your latest project. You've got a couple of Web sites for electronic parts, or maybe even the printed catalogs of electronic component distributors like Mouser. How do you start? Here's the way:

1. If it's an integrated circuit or some other component where you have a part number, go to www.findchips.com, and enter the part number into the search box. The more unique the part number, the better chance you'll have of finding what you want. For example, a search for *NE555* or *LM555* will likely find that 555 timer IC you want; entering just *555* is apt to put you on a wild goose chase.

2. If you get a match, go to the results page for any and all of the suppliers that have it. Double-check it's what you want. Compare prices.

3. If FindChips comes up empty, try your favorite online electronics retailer. All of them have a search box. Try entering the name and value of the part. Start with the specific part you want—*10 microfarad non-polarized capacitor.* If that doesn't work, simplify the search to just *non-polarized capacitor,* and then simply *capacitor.* The more generic the search, the more results you'll have to slog through.

4. If the search feature of the site doesn't provide answers, or if you're using a printed catalog, browse by category. For a Web site, locate the navigation links and click on them to narrow the search. For instance, you might start with *Capacitors,* then *Aluminum Electrolytic,* and finally *Non-polarized.* Review the value of the components that are listed. For a printed catalog, start with the index or table of contents, then browse the relevant pages.

5. Some electronics suppliers don't have printed catalogs, and their online search feature yields poor results. In that case you can use Google or another Web page search engine.

6. When all else fails, find the Contact page for the Web site, and ask directly if they have the part(s) you need.

Troubleshooting Tips, or How to Keep Things from Going *Worng!*

Where nothing can possibly *go worng!.*"... that was the catchphrase to the 1973 movie, *Westworld*. You know what happens: some programming glitch causes all the human-like robots in the Westworld theme park to go haywire. A particularly nasty gunslinger bot goes after the two heroes of the movie, shooting one dead (Josh Brolin's dad no less!), and chasing after the other until its face gets burned off with acid.

I never give my robots six-shooters for this very reason. I know sooner or later something won't work the way it should. It's bad enough to have to troubleshoot problems with batteries, let alone worry if my robot will shoot me in the back one dark night.

Troubleshooting is the step-by-step process of determining the root cause of a particular problem, whether it's electronic or mechanical. It's one of the core arts and sciences of robot building. In this appendix I provide some basic advice on how to troubleshoot your Arduino-based bots, gunfighter or otherwise.

Start with a Preflight Check

Ever heard the phrase "kicking the tires"? In the early days of the automobile, tires were made of a much thinner rubber than they are now. "Kicking" them before setting off on a trip would help a motorist know if the tires were ready for the road. Better to catch problems early. You can "kick the tires" of your Arduino robot by going through a kind of preflight checklist that demonstrates all systems ready.

Start with a visual inspection of the Arduino itself, its battery or other power connection, and all wiring. Note if anything's become disconnected or loose.

If you're programming the Arduino or powering it from your PC's USB port, be sure the cable is snugly inserted on both ends. USB cables and their jacks can become damaged during normal use—wires kink and plug contacts get bent or broken off.

If you suspect a problem has developed, but aren't sure of the cause or how extensive it is, upload a "signs-of-life" sketch to the Arduino to test basic operation. A good signs-of-life demonstration is the LED blink example sketch, or load up the similar one in the following listing. Upload it to the Arduino, then watch for the telltale flashing of the LED on pin 13 of the Arduino.

 Blinker

```
boolean ledState = false;

void setup() {
  pinMode(13, OUTPUT);
}

void loop() {
  digitalWrite(13, ledState);
  ledState = !ledState;
  delay(500);
}
```

Naturally, you can't use this method if your Arduino has something plugged into pin 13 that otherwise might alter the behavior of the LED. An example is a switch input where you've added your own pull-up or pull-down resistor. These external components will override the operation of the Arduino's built-in LED, causing it to either stay on all the time or not flash at all.

If you have a shield plugged into your Arduino you may wish to temporarily disconnect it before running any signs-of-life sketches. Removing the shield disconnects any electrical components that might be interfering with the operation of the Arduino. Parts on a shield can go bad or short-out, and those faults can prevent the Arduino from powering up or working correctly.

Systematic Approach to Development and Troubleshooting

Simple bots need only simple troubleshooting. They only have one or two features, and when something's wrong, the reason is usually obvious. But add complexity to the design—even something as innocuous as a separate set of batteries for the servo motors—and you're left wondering where to start in troubleshooting any problems. Layer upon layer of features and functionality create ever-increasing combinations of trouble hot spots.

As a system gets larger, working out the kinks becomes much more difficult. One of the best ways to test these kinds of systems is to break them down into smaller and more manageable units.

- Avoid the shock of a non-functional robot by building it in smaller pieces. Begin with the basic drivetrain—connect just the motors and battery to the Arduino (through a suitable H-bridge or other suitable circuit, as needed).

- If this part works you're ready to move on by building and testing each new capability separately. Add just one feature or capability at a time. Avoid the temptation to implement ultrasonic range finding, infrared detection, and music playing all at the same time. Each of these is a discrete unit; test each one to determine if that function is working properly.
- After building and testing the core blocks of your robot, methodically integrate them with the rest of your system. Start with those that are critical to the design of your robot, plus have the most impact on the operation of the Arduino—this allows you to more readily isolate problem areas up front. For example, music and sound playing can consume considerable processing time. Integrate it early with your robot's basic motor function. Add the other features (one at a time), like the ultrasonic and infrared sensors. By introducing the most demanding functions first, you can readily determine which additional function "breaks the camel's back," so to speak. You can then decide if another programming approach is preferred, or as needed which features to keep.

The same systematic techniques may be used to troubleshoot a fractious bot. Test each subsystem separately, ideally electrically isolated from the rest of the components. That may mean partially disassembling your robot and disconnecting parts you don't need. Circuitry on that music-playing shield may be interfering with some other piece on your robot; remove the shield completely; don't just rely on commenting out the music-playing code. Similarly, sensors and other components individually plugged into your robot's Arduino should all be removed, then replaced one at a time for testing.

Apart from easily resolvable problems like loose wiring or worn-out batteries, the interaction of the specific combination of circuitry on your robot is the main cause behind technical glitches. Some hardware may not be widely compatible with others, for example.

Or, circuits like cellular phone boards may consume copious amounts of current, which can cause constant or intermittent errors—in the case of a cell phone SMS shield, for example, the circuit may draw an amp or two when transmitting. If your robot's power supply can't handle the peak load, the sudden draw can cause the Arduino to momentarily shut down.

These so-called brownout events can be hard to track down, because the problem may only rear its ugly head when other current-consuming circuits are also added to the mix. Remove any single circuit, and your bot functions again. Add it back in, and the problem returns. On the surface this kind of mayhem can look like a fault in the individual circuits, when it's really just a matter of overtaxing the battery. In cases like these, using a bigger battery, or completely eliminating circuits that draw excessive currents, usually solves this problem.

Using the Serial Monitor to Debug Problems

Get into the habit of validating the operation of your robot and its Arduino processor by using the Serial Monitor window. This tool allows you to send messages from the Arduino back to your PC—these messages serve as a way to quickly and efficiently debug many aspects of your robot/Arduino, all at the same time. While you must take a few moments to set the debugging environment, it can save considerable time in the long run.

No doubt you've already used the Serial Monitor to display messages from the Arduino, so you already know the basic concepts. But just in case the idea is new to you, here's what's involved:

1. The hardware serial port is configured in the *setup()* function. Merely add:

```
Serial.begin(9600);
```

someplace within *setup()*, preferably near the beginning. The 9600 value is the baud rate, the data rate that the Arduino and PC will communicate with one another. This value is actually rather low, and both your computer and Arduino are capable of talking to one another at much faster speeds. However, 9600 baud is the most common rate used in the various Arduino examples, so it's the one used—with rare exception—throughout this book.

2. Debug messages are sent from the Arduino to the PC by using one or more *Serial.print* statements, like this:

```
Serial.print("ok");
```

which sends a simple "ok" to the PC. This statement sends just the text and no new line character. So when you view the debugging information in the Serial Monitor window, all the text will appear on the same line. If you'd like it to show up on different lines, use the variation

```
Serial.println("ok");
```

3. To actually see the debugging messages on your PC you need to open the Serial Monitor window, which is a feature that's built into the Arduino IDE software. After sketch download is complete, just click on the Serial Monitor icon, and the Serial Monitor window will appear. Verify that the Baud Rate setting in the lower corner of the window is the same as the *Serial.begin* statement used in your sketch. Otherwise, the Arduino and PC will not sync up with one another, and the Serial Monitor window will display nothing, or gibberish characters.

Use Serial Monitor debugging whenever you want to verify a part of your code is working as it should. The following listing provides a short working example.

Debugger

```
boolean ledState = false;

void setup() {
  Serial.begin (9600);   //Be sure to match in Serial Monitor
  delay(250);
  Serial.println("Setup complete!");
}

void loop() {
  digitalWrite(13, ledState);   // LED on or off
  Serial.print("LED on: ");     // LED state in Serial Monitor
  Serial.println(ledState);     // Shows 0 or 1
  ledState = !ledState;         // Toggle LED state
  delay(500);                   // Wait 1/2 second
}
```

Good places to insert a *Serial.print* statement are detailed in the following sections.

INSIDE THE SETUP() FUNCTION

This will verify that the Arduino is progressing to at least that point in its program. The *Serial* *.print* statement must appear after *Serial.begin*. If printing text immediately after *Serial.begin*, consider adding a quarter to half-second delay:

```
Serial.begin(9600);
delay (500); // Wait half a second
Serial.println("Setup!");
```

Inserting a debug statement inside *setup()* also helps you to verify that the Arduino runs the setup code just once. If you see the "Setup!" message repeat, it could mean your Arduino is being reset over and over again. This can occur, for instance, if the Arduino is not getting enough voltage to operate reliably.

AT THE START AND/OR END OF THE LOOP() FUNCTION

This will check that the Arduino is properly completing each loop. You can use static text, as in

```
Serial.println("Looping!");
```

or set up a more elaborate system where you indicate the number of loops that have elapsed so far.

```
int counter = 0;

void setup() {
  Serial.begin(9600);
}

void loop() {
  Serial.print ("Loop: ");
  Serial.println(counter++);
  // rest of code
}
```

Be careful when using debug statements in the *loop()* to prevent over-running the communications between the Arduino and your PC. If the loop doesn't contain programming statements that require at least half a second to execute, (temporarily) slow things by inserting a delay at the bottom of the loop:

```
delay(500);  // Wait half a second before proceeding
```

INSIDE A USER-DEFINED FUNCTION

This will verify that the function is being called. User-defined (your own) functions are a handy way to execute the same code whenever a certain event occurs. A good example is making the robot stop, back up, turn, and head off in a new direction whenever one of its touch sensor switches is activated. You can verify that the Arduino is reaching the user-defined function simply by adding a debugging statement as the first program line within it:

```
void rightBumperButton() {
  Serial.println("Right bumper activated!");
  // rest of code
}
```

AT PARTS OF THE SKETCH THAT USE LOOPS

This is where you want to be sure that program execution eventually continues beyond the loop. Suppose your robot stops and waits for a bumper switch to activate. The code is part of a loop, like so:

```
while (digitalRead(10)==0) {}
```

which means check the value of digital pin 10, and keep looping as long as the pin is LOW (0). All fine and good, but sometimes, programming errors, wiring mistakes, and other gremlins can cause your program to simply stop at this loop and never continue. You can check that the loop eventually exits just by placing a debugging statement immediately after:

```
while (digitalRead(10)==0) {}
Serial.println("While loop ended!");
```

WHENEVER YOU NEED TO PREVIEW A VALUE

You may do this for the result of a sensor, for example. Imagine you have an ultrasonic range-finding sensor, but aren't sure if you're experiencing problems because the value from the sensor isn't what you're expecting it to be. You can visually check any numeric or boolean value from within the sketch simply by sending that value to the Serial Monitor window. For instance:

```
int ping = getPing();
Serial.print("Distance is: ");
Serial.println (ping, DEC);
// rest of code

int getPing() {
  // Code to read ultrasonic sensor
}
```

In this abbreviated example, the code that does the actual reading from the ultrasonic sensor is contained in a user-defined function named getPing. You call that function with the line

```
int ping = getPing();
```

The numeric value that represents the reading from the sensor is contained in the *ping* variable, which you can use in a debugging statement. Note the DEC qualifier—it ensures that the number held in the *ping* variable is expressed as a decimal value. The Arduino supports other qualifiers as well, such as HEX and BIN, which display the value as a hexadecimal (base 16) or binary number, respectively. Check the Arduino reference at the arduino.cc Web site for additional options. By far you'll use the DEC qualifier the most.

The Arduino's Serial Monitor supports a couple of control characters—also called escape sequences—that can come in handy when you want to display several debugging values at once. All control characters begin with a \ (backslash). Probably the most useful control character is \t, which inserts a tab in the print statement line:

```
Serial.print("Some text here");
Serial.print("\t");
Serial.println("Some more text");
```

You don't need to emplace the \t control character in its own *Serial.print* statement when using just literal strings, like that shown earlier. But you need to if you want to mix literal strings plus numeric values from variables:

```
Serial.print("Sensors reading\t");  // Explanatory text plus tab
Serial.println(mySensor, DEC);      // Value from mySensor variable
```

or

```
Serial.print("Sensor 1 is: ");    // Explanatory text
Serial.print(Sensor1, DEC);       // Value in Sensor1 variable
Serial.print("\t");               // Tab
Serial.print("Sensor 2 is: ");    // Repeat for sensor 2
Serial.println(Sensor2, DEC);
```

Some Common Quandaries, and How to Fix Them

Troubleshooting is really just a systematic way of analyzing a problem, then considering the one or more likely causes of that problem. From such beginnings you can then investigate each theory about why things aren't working correctly. With a little bit of experience, odds are you'll find the failure within a few attempts.

Here are some of the most common problems that beset Arduino-powered bots, and the most common ways to fix things.

NO POWER, NO LIGHTS, IT'S AS PRIMITIVE AS CAN BE

A completely non-functional robot strongly suggests a problem with the power supply. If having problems with your robot, start first with its brain, the Arduino. If the Arduino's power light is not on, the most likely reason is that it's not getting power. Start with these checks:

- *Weak or dead batteries.* Recharge or replace.
- *Bad USB connection, if powered from your PC.* Check the connection to see if the cable is merely loose. Try a different cable if one is available.
- *Faulty power switch or wiring.* Small switches can go bad, especially if struck hard during a collision. Inspect and replaced, as needed. While looking see if the wires from the battery have come loose. Check the power connection (barrel plug) as it goes to the Arduino.

ERRATIC BEHAVIOR

Your robot is working; it just doesn't work all the time, or it occasionally behaves in bizarre ways. The most common reasons are:

- *The battery charge is running low.* Replace, recharge, or increase the battery capacity (larger battery and/or higher voltage, as needed).
- *Bad wiring connection.* A broken or loose wire anywhere, especially ground related, can cause all manner of problems. If using solid conductor wire, check with your meter while gently moving the wire side to side. You'll get intermittent continuity if the wire is broken inside the insulation.
- *Wet or dirty contacts.* Water and dirt can cause intermittent connections. Wipe clean, dry, and try again.

LIGHTS ON, NO SERIAL DEBUG MESSAGES

Suppose the Arduino's power LED is glowing, but nothing seems to happen. What's more, you've used debugging messages in your sketch, and these messages are not appearing when you open the Serial Monitor window.

- *The USB cable is bad, or the wrong serial port is selected.* This error can occur if you've previously uploaded your sketch and have reconnected the Arduino to your computer. Verify the integrity of the cable, and make sure the proper serial port is selected. (If you have more than one Arduino, each one may be associated with a different serial port. Be sure you've picked the right one.)
- *Wrong baud rate selected in the Serial Monitor window.* Be sure it matches the baud rate specified in your sketch.
- *Serial communications out of sync.* Sometimes your Arduino and PC fall out of sync, and they will not re-sync unless you (a) reset the Arduino and/or (b) reset the Arduino IDE (close then reopen it).

SERIAL WINDOW DISPLAYS GARBAGE

When you open the Serial Monitor window to look at the debugging messages sent from the Arduino you get garbage text instead.

- *Wrong baud rate selected in the Serial Monitor window.* Be sure it matches the baud rate specified in your sketch.
- *The USB cable is bad.* Verify the integrity of the cable, or try a different one.

DEBUGGING MESSAGES REPEAT WHEN THEY'RE NOT SUPPOSED TO

You've added a debugging message to the *setup()* function of your sketch. It should display once in the Serial Monitor window, but instead it keeps repeating.

- *Arduino is resetting by itself.* Look for causes of power brownout—the voltage to the Arduino is falling below the minimum required. When this occurs the sketch spontaneously restarts.

- *Arduino is resetting via the serial port.* On most versions of the Arduino, plugging or unplugging the USB cable causes the serial port to reset, which in turn causes the sketch to reload from the beginning. Look for causes of spontaneous reset of the serial port, which can occur, for example, if the cable or plug connection is bad, or if the Arduino is drawing too much current from your PC's USB port.

DATA RESULTS ARE WRONG/UNEXPECTED

You're getting debugging messages, plus the data shown is wrong or unexpected.

- *The sensor or other device is misconfigured.* If you're checking the results of a sensor, make sure it's properly configured and programmed. For example, an ultrasonic range-finder sensor may be programmed to return distance in inches or centimeters (or raw microsecond delay of the echoes). The data will appear wrong if you're expecting it in a different format.
- *Sensor or other data in the wrong format.* The *Serial.print* statement can use an optional data formatting argument, specifying if the data is to be shown in DECimal, BINary, HEXadecimal, or other format. Be sure to use the proper formatting argument based on the type of data value.
- *Programming error.* Though we try to avoid bugs in code, sometimes they creep in and muddle the results. This is what debugging is for—catching those mistakes so they can be fixed. Carefully analyze your programming code, and look for math mistakes, incorrect assumptions, and other errors that can cause the invalid data. For example, did you really mean to add those two numbers together, or multiply them?

SKETCH (PROGRAM) WON'T COMPILE

The most annoying glitch when using the Arduino is some type of error that prevents the sketch from compiling and uploading. Though there are many reasons this can occur—including an improperly installed IDE software package—following are among the most common reasons.

- *Syntax error in code.* Check for proper spelling and capitalization of all Arduino keywords. Remember, it's digitalWrite, not DigitalWrite, digitalwrite, digitalRight, or some other variation. Also check for missing semicolons, parentheses, braces, and other typical errors.
- *Incompatible version of IDE software.* The Arduino IDE software has gone through many changes over the years, where some changes have "broken" existing sketch code. Be sure you're using the correct IDE version for your sketch. Unless there is a specific reason otherwise, it's best to keep your Arduino IDE up to date with the latest stable release. However, you may keep previous iterations installed on your computer, should you need to compile your sketch with an earlier version.
- *Missing or incorrect library.* Sketches that must be compiled with one or more libraries require those libraries in specific locations—either the main *libraries* folder in the Arduino IDE program directory, or the *libraries* folder within your sketchbook directory. If installing a new library be sure to exit and restart the IDE before attempting to use it.

EVERYTHING WORKS EXCEPT THE MOTORS

Your Arduino appears to be functioning normally, but your robot's motors aren't turning.

- *Not enough current to power the motors.* Motors can draw a lot of current from your robot's batteries. If the batteries don't deliver enough current, the motors may not turn, or they may turn only slowly. Beef up the batteries to ensure proper motor current levels.
- *Separate motor power disconnected, discharged.* If your robot uses a separate power source for the Arduino and its motors, the motor power may have become disconnected, or the motor batteries may be discharged.
- *Shared ground disconnected.* Robots that use separate battery supplies for the motors need to have the ground of the two power sources connected. (There are exceptions to this, such as when the control board for the motor is interfaced to the Arduino using an opto-isolator.) Otherwise, the robot may not operate reliably, or at all. So double-check the common ground connection between the two power sources.

Index

Notes

Notes